*To the Sustainability of Earth*

Global Climate Change, Environment and Energy:
Global Challenges and Opportunities to Global Stability,
Edited by Filiz Katman

This book first published 2014

Cambridge Scholars Publishing

12 Back Chapman Street, Newcastle upon Tyne, NE6 2XX, UK

British Library Cataloguing in Publication Data
A catalogue record for this book is available from the British Library



ISBN (10): 1-4438-6693-8, ISBN (13): 978-1-4438-6693-4

# Global Climate Change, Environment and Energy: Global Challenges and Opportunities to Global Stability

Edited by

Filiz Katman

CAMBRIDGE
SCHOLARS
PUBLISHING

# Global Climate Change, Environment and Energy

# TABLE OF CONTENTS

# LIST OF TABLES

# LIST OF FIGURES

# PREFACE

## FILIZ KATMAN[1]

Climate change presents a fundamental challenge for global sustainable development. Global climate change is both an environmental and sustainable development issue. On 17 April 2007, for the first time, climate change was raised as an issue at the United Nations Security Council. The Security Council held its first-ever open debate exploring the relationship between energy, security and climate. The UN Security Council stressed the need to consider economic, social and political factors in maintaining international peace and security at its 6479th Meeting, on 11 February 2011. The effects of climate change were addressed as factors that have an increasingly serious impact on development and security (SC/10172). The European Council has drawn attention to the impact of climate change on international security and, in June 2007, invited the High Representative and the European Commission to present a joint report to the European Council. In March 2008, Higher Representative Javier Solana and the European Commission presented a report to the European Council entitled *Climate Change and International Security,* analysing the consequences of the adverse effects of climate change on international security (S113/08). Global climate change is the combination of natural climate change and climate change due to human impacts that, directly or indirectly, damages global atmosphere components observed over a comparable time period. It has both a direct and an indirect impact on earth. On the other side of the coin, it brings opportunities such as the possibility of a green economy.

This book[2] aims to cover both sides of the coin in providing comprehensive information concerning climate change, environment and

[1] Assistant Professor, Department of Political Science and International Relations, Faculty of Economics and Administrative Sciences; Executive Board President Energy Politics and Markets Research Center, Istanbul Aydin University, TURKEY, filizkatman@aydin.edu.tr

[2] The author(s) of each article bear full responsibility for the scientific content of their papers and compliance with scientific and ethical rules. The views/ figures/tables expressed by the authors are entirely their own, and the editors shall

energy. It comprises selected articles discussing these issues from a multidimensional perspective and elaborating upon their interdependence. The articles are based upon contributions to an International Symposium on Global Climate Change, Environment and Energy-Global Challenges & Opportunities for Sustainability, organized by the Istanbul Aydin University Energy Politics and Markets Research Centre on 25 April 2011, held at Istanbul Aydin University. The aim of the conference, and this volume, is to fulfil the responsibility owed to the earth and to humanity; and it would not be available without generous support of Prof. Dr. Hasan Saygın.

---

bear no responsibility for content or for copyright violation issues. It is the responsibility of the authors, not the editors, to determine whether disclosure of their material requires the prior consent of other parties and, if so, to obtain it.

# INTRODUCTION

# GLOBAL CLIMATE CHANGE, ENVIRONMENT AND ENERGY: GLOBAL CHALLENGES AND OPPORTUNITIES TO GLOBAL STABILITY

FILIZ KATMAN[1]

Global climate change refers to an increase in the temperature of the atmosphere and the oceans. Some gases, such as carbon dioxide ($CO_2$), Methane ($CH_4$), nitrous oxide ($N_2O$) and hydro fluorocarbons (HFCs) lock some of the thermal energy disseminated through radiation from sun and earth, leading to an increase in the temperature of the atmosphere. This is called the greenhouse effect, and gases that create this effect are called greenhouse gases. The greenhouse effect is a situation present in nature. The problem of global warming originates in the intensification, to dangerous levels, of the level of greenhouse gases in the atmosphere, and originates in human activity. Although carbon resources, primarily $CO_2$, are at a stable level, the strongest temperature catcher, greenhouse gas emissions in the atmosphere, have increased rapidly due to modern human activities accompanying production-consumption processes. Global warming due to increasing greenhouse gas concentrations in the atmosphere has reached very dangerous levels. Today, there is an academic consensus that 90% of global warming originates in human activities (Saygin, 2007: 38-39). Global climate change is the most serious consequence of global warming. The rise in the temperature of the atmosphere and the oceans due to increasing greenhouse gas emissions is

[1] Assistant Professor, Department of Political Science and International Relations, Faculty of Economics and Administrative Sciences; Executive Board President, Energy Politics and Markets Research Center, Istanbul Aydin University, TURKEY, filizkatman@aydin.edu.tr

the combined effect of natural climate change and climate change due to human impact. This damages global atmosphere components and has both direct and indirect impacts on the earth.

Changes in the global climate may affect human life and the ecology of the planet in several ways: water floods originating from increased sea levels due to melting ice; loss of soil on the shore; clean water being mixed with sea water; other issues relating to drinkable water, fires, and droughts originating from the excessive vaporisation caused by temperature increase; an approximate 20% decrease in lake and river waters; the depletion of some types of plants and animals due to such changes; epidemic diseases caused by the mutation in viruses due to excessive temperature increase, and overcapacity in local and global carriage due to possible migration waves. These are just some of the possible problems that may arise in the near future (Saygin, 2007: 38-39). Thus, combating global warming and climate change originating from global warming is very high on the agenda of many countries, primarily developed countries. As a result, the term ‘global warming’ has entered into everyday terminology.

Risks associated with global climate change are of enormous significance to the future of international society. It is a reality accepted by international society that the problem of climate change will have a serious impact upon future energy systems. It was in the nineteenth century that international agreements concerning the management of environmental issues first emerged. The River Commissions and The International Maritime Organisation are the first examples. 1960 was the year in which environmental concern grew rapidly, particularly in developed countries. Oil pollution at sea, the widespread use of DDTs, radioactive fall-out, transboundary air pollution, and acid rain were the principal global concerns that led to the organization of the 1972 Stockholm Conference. This was a UN Conference on the Human Environment which investigated the prospects for an international framework and a coordinated effort to fight pollution and other such environmental issues. Its outcomes (principles, institutions, and programs) represented an important beginning.

The primary concern was the responsibility of states to cooperate in the effort to manage global common resources and environmental resources. The main clash arose with regard to the share of responsibility for environmental issues. Developing countries argued that they shouldered less responsibility than developed countries and that sustainable development should be the main focus of such efforts. In a milestone development, it was at this meeting that the relationship between environment and development with regard to the North-South axis formally emerged. It involved the participation of several non-governmental environmental organizations, which put political pressure on the participants. This conference was the first of several further agreements on the environment held during the 1970s and 1980s. Twenty years later, a mature environmental politics was created when a UN Environment Program was established as a key institution coordinating environment-related activities. This also led to the establishment of green movements and green parties. In 1989, the so-called Rio Conference was organised by the UN General Assembly—the "Earth Summit", UN Conference on Environment and Development (UNCED)—in order to focus systematically on sustainable development.

By the 1990s, at least 120 multilateral environmental agreements and hundreds of bilateral ones were signed (Owen, 1997: 324). After this large number of environment agreements, the next step in the development of the institutionalised approach was the formation of an international regime on the environment, composed of several phases, including agenda formation, negotiation and decision-making, implementation and further development. Ozone layer depletion is an important and effective international environmental regime based on The Montreal Protocol of 1987. In terms of agenda formation, the first steps towards this protocol were taken during the 1970s.

The Rio Conference in 1992 succeeded in establishing three new conventions on limiting climate change, preserving biodiversity and combating desertification: however, the last two regimes stalled in 1996. In terms of agenda-setting, The Intergovernmental Panel on Climate Change (IPCC), composed of scientists under the auspices of the UN Environment Program and the World Meteorological Organisation, examined the impact of anthropogenic greenhouse gas emissions. The report prepared by the IPCC in 1990 (by the representatives of 137 countries at the Second World Climate Conference in Geneva) warned the international community of the urgency of the situation, and led to the signing of the Framework Convention on Climate Change (FCCC) at the Rio Conference. Furthermore, concern for natural habitats and species led to the Convention on Biological Diversity in 1991, based on a range of expert reports. As regards desertification, concern for deforestation (particularly of the tropical rainforests) led to the Convention to Combat Desertification, which was widely supported by industrialised countries despite the fact that the issue was of primary concern to African countries. In addition, the Rio Declaration contributed to the agenda-setting for the Rio Conference, in setting out a list of agreed principles called Agenda 21.

The United Nations Framework Convention on Climate Change-UNFCCC in Rio de Janeiro in 1992 was the ground of global efforts to combat global warming. It aimed to prevent dangerous activities leading to climate change. The parties involved agreed to take mutual but proportionate responsibility. It was agreed, firstly, that developed countries had the largest share of greenhouse gas emissions; secondly, that emissions per person in developing countries were still low; thirdly, that greenhouse gas emissions in the latter countries would continue to increase in order to complete the process of economic and social development.

Framework Conventions were subject to ratification by a minimum number of parties, as defined in the treaty, and these were enforced very rapidly (within 2 years). Conferences of parties were organised, and the first of these was held in Berlin in 1995. Its purpose was to negotiate a limitation on greenhouse gas emissions. Almost all the OECD countries and the European Union member states agreed to stabilise their emissions at 1990 levels by 2000, but progress towards achieving this goal was not sustained. One critical impact of climate change has been the increase in sea levels resulting in inundation, and this led the Alliance of Small Island States to advocate for a reduce emissions by 20 percent in 2005. This move, however, was met with strong resistance by the OPEC countries (Greene, 1997: 324).

The EU and Western countries supported the reduction of emissions by 5-10 percent in 2010, but several other developed countries rejected this proposal. Reports on national greenhouse gas emissions and limitation measures were reviewed while negotiations were ongoing. This reflects the fact that an effective climate change regime requires long-term processes, involving development and implementation of the international environmental regime under an umbrella framework made up of UNCED agreements.

The most crucial milestone in combating climate change is the Kyoto Protocol, which builds on the principles discussed above. Obviously, the primary goal of the Kyoto Protocol is to stabilise greenhouse gas in the atmosphere at a level that prevents anthropogenic climate change. After long negotiations, the Protocol was presented in Kyoto City in Japan, on 11 December 1997. For the purpose of legal enforcement, the condition was the acceptance of the obligations of the Protocol by 55% of the countries listed in Appendix I of the UN Framework Convention, since 50% of emissions in 1990 originated from these countries (Saygin, 2007: 38-39). Thus, the Protocol was well developed when the US, under the George W. Bush administration, announced the withdrawal of US support in 2001, on the grounds that exempting developing countries from emission restrictions would prevent the success of the Protocol and damage the US economy. Australia, the first country on the list and with the highest level of emission per person, had initially led the negotiations, but then changed their position and refused to sign the Protocol at the last minute. Thus, US withdrawal from the Protocol had a dramatic effect, and led to serious suspicions and arguments about the feasibility of the Protocol. In order to analyse these developments, a summit was organized in Bonn in June 2001, with 180 countries in attendance. It was agreed that the Kyoto Protocol was on the brink of collapse, and that precautions should be taken to prevent this. The EU took the lead in designing strategies to ensure the sustainability of the Protocol without the participation of the US.

In 1997, the Kyoto Protocol on greenhouse gas emission abatement imposed an obligation on developed countries to decrease their dissemination to 5.2% below 1990s levels, over the period 2008-2012 (Saygin, 2005: 27). Three market-based mechanisms were formed in order to achieve these goals: (i) Emission trading (among Annex I countries) and international emission trading. This gave countries with greenhouse gas emission levels below the determined limits the right to sell their remaining emission rights. (ii) Joint Implementation (JI) allowed project-level emissions trading between Annex I countries, through funding opportunities for such projects, deducted from determined levels. (iii) The Clean Development Mechanism allowed Annex I countries to deduct from their total emission targets, in case of investing in project-level emissions through funding (as in JI).

In order to be implemented, this agreement was subject to the approval of at least 55 countries contributing to $CO_2$ emissions of developed countries in 1990. The withdrawal of the US from the agreement was the primary shadow cast over the original vision of the Protocol. The US, under the Bush administration, had argued that the economic cost was too high and that it was unacceptable to exempt developing countries from the primary emission targets. The US officially announced its withdrawal in March 2001. This influenced the environmental and cost efficiency of the Protocol, and many questioned the applicability of the Kyoto Protocol in the absence of US participation. In June 2001, the possible collapse of the Kyoto Global Warming Agreement within 10 years was identified, during negotiations between representatives of 180 countries in Bonn. The decision was taken to prevent this collapse by any available means. The EU took the lead, forming strategies in order to secure the sustainability of the Protocol without US participation.

On the other hand, the participation of former Soviet Union countries (primarily Russia and Ukraine) and Eastern European countries (Japan) was crucial in achieving the required majority for the enforcement of the Protocol. Strategies for preventing a possible collapse, subject to the participation of these countries, led to the granting of privileges that risked diminishing the primary impact of the Protocol. Thus, the primary goals of the Kyoto Protocol were reviewed at the meetings held first in Bonn, then in Marrakesh. New options in Bonn, such as meeting one part of the limitations on $CO_2$ emissions through $CO_2$ stored in forests and farming areas materialised, and these were finalised in Marrakesh. The primary outcome of the Marrakesh Agreement was the addition of $CO_2$ sinks, a substantial number of $CO_2$ credits called sink credits (for countries such as Australia, Canada, New Zealand, Japan and Russia), to be deducted from the targeted total amount for emission abatement for forests and farming areas storing greenhouse gases. Importing $CO_2$ sink credits is the most crucial step in softening and weakening the impact of the primary Kyoto Protocol, and reflects the latest version of the protocol, after the Marrakesh meeting. This newer version of the Protocol tends towards meeting all abatement obligations through emission trading, and creates a space between permitted amounts and emissions for sale; it does not project the possibilities for contradiction into the essence or the principles of the Protocol. In fact, Russia played a critical role in the negotiations, as a key country in the enforcement of the Protocol, and it was 'a success' that doubling $CO_2$ absorption credit via the Marrakesh Agreement went through at Russia's request.

In December 2011, a summit organised in different countries and cities was held in Durban City in South Africa. At this summit, it was decided to determine the greenhouse gas abatement and the financial and technological obligations of developed countries after the first obligation period of the Kyoto Protocol. It was also agreed to ensure the enforcement of the Green Climate Fund in 2013.

The last summit was held in Qatar, the most problematic country in terms of $CO_2$ emissions. Several developed countries, including EU member countries and Australia, agreed to reduce their $CO_2$ emissions until 2020. However, the non-participation of the USA, China and India, the most atmosphere polluting countries, made the Protocol meaningless. Ironically, the next summit (the $19^{th}$ Conference of the Parties to the United Nations Framework Convention on Climate Change) will be convened in Warsaw, Poland, a nation with a heavy dependence on coal that prevents it from meeting the strict emission targets of the EU.

In the lead up to the 19th Conference, two critical steps have been taken by the US, in terms of contributing efforts to the problem of climate change. In July 2013, President Obama of the US and President Xi of China reached an important agreement to jointly phase down production and consumption of hydro fluorocarbons, and it was agreed to take further steps in the coming months. In a speech at the White House in 2013, President Obama (the President of the world's largest economy and second-largest carbon emitter) announced measures to reduce carbon pollution and to protect people from climate change. He announced his new climate change policy on June 25, 2013, at Georgetown University. In that speech, he emphasised a low-carbon and clean energy economy as an engine of growth for decades to come, based on using less dirty energy, transitioning to cleaner sources of energy, and wasting less energy. He also outlined some concrete steps, such as the following: doubling energy from wind and sun to power more than 6 million homes by 2020; installing 3 gigawatts of renewable power on the bases of the Department of Defence—the biggest energy consumer in America, which generates about the same amount of electricity each year as would be generated by burning 3 million tons of coal; calling upon Congress to end tax breaks for big oil companies; investing in clean-energy companies; wasting less energy in cars, homes, and businesses, via manufacturing and installing smarter lights and windows and sensors and appliances; setting new energy standards for appliances, and providing 20 percent of the federal government's electricity supply from renewable resources within seven years. It should be noted that while in 2012, American emissions of $CO_2$ from fossil fuels fell by 200 million tonnes (Mt) to levels last seen in the mid-1990s (due to the transition from coal power to natural gas and renewable energy), in the same year global carbon pollution rose to a record high (by 1.4 percent) to 31.6 billion tonnes (gigatonnes [Gt]). This is the second smallest annual increase since 2003 after 2009 (which was due to the global recession) (IEA, 2012). President Obama's speech created an expectation for further steps, taken globally at the 19th Conference on Climate Change.

In this book, global climate change (a critical issue in world politics), will be discussed based on an approach which stresses the interdependence between environment and energy. The issue of the global commons requires an interdisciplinary and multidisciplinary approach. The book identifies the challenges and policy recommendations required to overcome these challenges. This makes the book unique.

In the first chapter, issues of global climate, environment and energy are discussed by the authors with respect to global challenges and opportunities for global stability, in three articles with a primary focus on the security aspect of global climate change. In the article entitled "Analysis of Global Climate Change Impacts on Security" by Filiz Katman, global climate change is presented as a contemporary threat to security, and its impacts on security in general terms are analysed. In the article entitled "Paradigm Shift in Millennium Development Goals Management through Global Climate" by Aylin Met and Taner Altunok, the circumstances surrounding the Millennium Development Goals and climate change are explained and the necessity and the applicability of paradigm change are discussed.

The article entitled "Seeking Security in Changing Climate" by Argiro Vasilaki takes the findings of the official documents of the European Union and the United Nations as its starting point, and analyses the social and human consequences of global climate change, with particular attention to the risks of instability and social unrest.

Chapter Two is dedicated to a discussion of the impacts of Global Climate Change on the Environment, with six articles by scholars, with a multidisciplinary approach. The article entitled "A New Ethical Approach to Environmental Protection: Ecocentrism" by Kıvılcım Akkoyunlu Ertan and Birol Ertan aims to explore the role of ecocentrism (the new ethical approach that has to be a guide for decision-makers and citizens in environmental protection) in preventing environmental problems, and its effects in the process of energy production and consumption.

In the article entitled "Voluntary Carbon Trade: A New Livelihood Model for Low Income Families" by İlker Met and Taner Altunok, practices of carbon market for low income families and a resulting model for Turkey are presented. A way of saving low income families (the most vulnerable to the effects of climate change) is recommended, as well as an important step towards mitigating the negative effects of climate change, by helping the carbon market to liberate itself from the monopoly of large projects and making it available to the community.

The article entitled "Analysis of Turkish Environmental Energy Policies in a Regional Framework", by Çiğdem Üstün, demonstrates the main issues that Turkey focuses on in its (foreign and domestic) environmental policies and in the security of energy supplies, using the findings of in-depth interviews conducted in Turkey and the regional countries, i.e. Greece, Moldova, Bulgaria, Ukraine, Georgia, and Azerbaijan.

The article entitled "European Union Environmental Directives and a Practice on Business" by Meral Erol Fidan presents the existing situation in Turkey, and measures the level of businesses' knowledge about environment and waste management in the adaptation process to the EU.

The article entitled "Wind Speed and Energy Characteristics based on Statistical Analyses" by Zafer Aslan and Z. Nevin Çağlar covers some statistical analyses of wind speed data, recorded in the Northwestern part of Turkey, Şile (Istanbul); monthly, seasonal and annual wind speeds are analysed statistically.

Chapter Three discusses the impacts of global climate change on energy. The article entitled "Sustainable Development and Turkey's Biomass Energy Potential" by Coşkun Karaca and M. Mustafa Erdoğdu focuses on the level of Turkey's biomass energy potential, and discusses ways of making sufficient use of this potential. It seeks answers to two questions in particular: (1) to what extent can the quality of the environment be improved by the use of biomass energy? (2) What changes occur in economic variables such as foreign trade, employment, and balance of payments when the consumption of fossil fuel is reduced by the use of biomass energy?

The article entitled "Energy and Climate Change Policies of Russia and China" by Çağla Gül Yesevi and Ceren Uysal Oğuz aims to analyse the meaning and dimensions of global climate change, energy and the environmental policies of Russia and China. The article entitled "The Relationship between Natural Resources and Security (with an Emphasis on Energy Security)" by Hakan Korhan addresses the place of energy security in national security definitions, the concept of energy security in terms of different purpose typologies of foreign policy, and the question of what kinds of foreign policy outputs such components might create.

## References

Greene, Owen (1997). "Environmental Issues" in The Globalization of World Politics, John Baylis and Steve Smith (Eds), Oxford: Oxford University Press.

IEA (2012). World Energy Outlook Report, http://www.worldenergyoutlook.org/media/weowebsite/2013/energyclimatemap/RedrawingEnergyClimateMap.pdf

Saygin, Hasan. (2007). "Enerji Koridoru: İklim Değişikliği ile Mücadelede Kyoto Protokolü" (in Turkish), Tekstil İşveren, Türkiye Tekstil Sanayii İşverenleri Sendikası Aylık Dergisi, Sayı: 334, Sayfa: 38–39.

—. (2005). "Enerjipolitik: Başlangıçtan Günümüze Kyoto Protokolü", Enerji, 10: 7, p. 27.

White House (2013). "Remarks by the President on Climate Change", 25 June 2013 http://www.whitehouse.gov/the-press-office/2013/06/25/remarks-president-climate-change

# Chapter One:

# Global Climate Change, Environment and Energy: Global Challenges and Opportunities to Global Stability

# ANALYSIS OF GLOBAL CLIMATE CHANGE IMPACTS ON SECURITY

FILIZ KATMAN[1]

## 1. Introduction

The needs of human beings change over time, but security has always been part of the most fundamental level of the hierarchy of needs. Security is a subjective feeling, determined by threat perception. If something normal is perceived as an existential threat—it might be presented as such a threat, even it is not a real one—then this process is called securitization. In desecuritization, a perceived threat is transformed into a normal issue. Whilst the means of providing and sustaining security are developing, threats to security are in the process of being transformed. September 11 proved that we live in an insecure world and that the source, time and place of a threat cannot be known before it is realized, so that the means of providing our security might not be appropriate. In such an interconnected world, a problem in one part of the world can affect a distant country: the problem becomes global. From this perspective, it can be argued that destabilising regions, concepts and insecurity (especially in strategic and important regions) are the objects of the process of securitisation, of desecuritization and, ultimately, of transformation (Buzan et al., 1998). Global climate change is the latest object of securitisation.

## 2. Global Climate Change: Crisis of the Earth

It was in 1826 that Joseph Fourier determined the role of the atmosphere on average world temperature; then, in 1896, Svante Arrhenius noted that coal used as an energy resource by a rapidly growing industry produced carbon gas (Denhez, 2007: 8). Global climate change is both an

[1] Assistant Professor, Department of Political Science and International Relations, Faculty of Economics and Administrative Sciences; Executive Board President, Energy Politics and Markets Research Center, Istanbul Aydin University, TURKEY, filizkatman@aydin.edu.tr

environmental and sustainable development issue. The word 'climate', meaning angle, is referred to as *klimatis* in Latin and as *klima* in Greek (Denhez, 2007: 11). Global climate change impacts are projected over the next 20-30 years, mainly due to greenhouse gases. In order to determine climate conditions, average values of air condition are observed over long periods of time. If the scale of the changes in climate is global, and impacts are significant and local, this means that climate change has occurred. On the other hand, global climate change reflects both natural climate change and also climate change due to the (direct or indirect) damage caused by the human impact on global atmosphere components, observed over a comparable time period.

At present, $CO_2$ emissions play a critical role. If internationally binding agreements are not reached on limiting $CO_2$ emissions, as well as on other greenhouse gas emissions, the upper limit of a 2-degree temperature rise will be exceeded. It is projected that the average global temperature will rise by 3.6 degrees Celsius by 2035 (IEA, 2012) with a domino effect, and in the long run, by 5.3 degrees Celsius; this is considered destructive for the world.

There are several indicators of global climate change, such as $CO_2$ increase per year, average global earth temperature increase, increase in sea level, and global warming. Since 1850, concentrations have increased by approximately 35% (Houghton, 2007: 319). The timing of this increase is coincident with the rising carbon emissions resulting from fossil fuel combustion and land-use change. Research shows the $CO_2$ increase per year has developed as follows:

- 1850 (approximately the start of the industrial revolution-2005: 0.65 ppm
- 1995-2005: 1.90 ppm (more than double)

The average global earth temperature increase was 0.74°C in 1906-2005, according to the Intergovernmental Panel on Climate Change-IPCC (2007). 1998 was the hottest year, with a 0.58°C anomaly. 2005 was the second hottest year, with a 0.49°C anomaly. After the 1990s, fluctuations occurred in the amount of rainfall. Another indicator is the increase in sea level. In the 20th century, the increase was 17 cm; the average yearly increase was as follows:

- 1961-2003: 1.8 mm (1.3-2.3 mm)
- 1993-2003: 3.1 mm (2.4 mm-3.8 mm)

Global warming is around 0.74°C. IPCC-Projections on temperature for 2000-2100 have two scenarios, optimistic and pessimistic:

- Optimistic scenario: 1.8 °C increase, $CO_2$: 600 ppm
- Pessimistic scenario: 4.0 °C increase, $CO_2$: 1550 ppm

Sea level increase in 2000-2100 is expected to be:

- Optimistic scenario: 0.18-0.38 m increase
- Pessimistic scenario: 0.26-0.59 m increase

The IPCC (2007) prepared scenarios on global climate change with no decrease in greenhouse gas emissions. According to scenario A1, the world population will be 7 billion with a rapid economic growth in 2100. In scenario A2, the world population will be 15 billion. In scenario B1, population growth is the same as in scenario A1. Economic structures, service, and the knowledge economy will diminish, due to a decrease in the financial economy, as will the use of clean energy and energy saving technologies. In scenario B2, despite diminishing growth, world population will be around 10 billion but with a lower rate of population growth than in scenario A2. The search for local solutions and economic development will be slower than in the A1 and B1 scenarios.

Based on such data, in order to prevent emissions reaching twice the level reached in the period prior to the industrial age, and notwithstanding countries' aspirations for economic development, emission in 2100 should diminish to half the 1990 levels. In order to keep the level to 550 ppm in 2100, an individual footprint should be 500 kg, which means 25% of a French footprint and 8% of an American footprint in 2005. If 550 ppm is realised, the average temperature increase in the atmosphere could remain at a level of 2-4.5 °C. If not, ppm levels will rise to 1000-1200 ppm, meaning that the average temperature increase will be 3.7-9.4 °C. Scenarios for 2100 show that under the most optimistic scenario, the increase will be 1.5-2.8 °C, 2.3-4.1 °C in the medium scenario, and 3.5-5.8 °C in the most pessimistic scenario.

## 3. Impacts of Global Climate Change on Security

Global climate change is detected by means of the previously mentioned indicators. Such changes have impacts which can be categorised as either direct or indirect. Direct impacts reflect the consequences of global climate change on human beings and the world. In contrast, indirect impacts reflect the secondary consequences of global climate change due to deteriorating conditions and the situations that arise as a result.

In 1896, Arrhenius projected that the average increase in global earth temperature would be 5-6 °C, due to a doubling in the amount of carbon gases in the atmosphere. However, according to IPCC (2007), the increase will be 1.5-1.8°C. As a result of this increase, water cycles will accelerate, ocean drifts and air cycles will change, and great biogeographical regions will be disrupted. Such changes in biogeography maps will challenge geopolitical conditions, and may lead to millions of 'climate migrants' facing hunger, drought, and distorted living conditions. Climate determines air, environment, activities, feelings and mood.

The Global Warming Atlas (Denhez, 2007) reviews the impacts of such changes on the environment, and provides a case of chain reaction. The ecologic environment is defined by temperature, amount of rain, duration of the seasons, and flora and nutrition conditions strictly defined for each species. In case of changes to any of these, a species will adapt, given physiological capacity and time. If not, the search for a new environment begins and, if none is found, the species disappears. The environment vacated by the species is then open to adaptive species. On the other hand, everything in nature is connected, so when one species changes its environment, this has an impact upon reproductive chains and hunting ranges. As a result, other species are also affected. This is called the domino effect. Researchers have observed that there has been a general move in species from the south to the north, due to global climate change.

The parameters of global climate change are listed as temperature, humidity, pressure, speed, etc. Changes in such parameters have a range of impacts, including the following: a decreasing frequency and increasing density of storms towards the north; an increase in floods, due to the increasingly unstable atmosphere; changes to the shores and continents, and the environment and types of plants and animals; changes to human beings in general. It is argued that the increasing levels of gases causing the greenhouse effect will lead to rain and warming, and that regional weather patterns will be distorted. There will be more rain, but less rain in some dry regions. It is expected that rain volume will increase 2-7%, due to rapid vaporisation-condensation cycles, leading to heavy and intermittent rainfall (Denhez, 2007).

As a chain reaction, sea levels will also increase, but it is argued that this will not be due to the melting of ice. The result of the increase in average world temperature will be expansion and increase in water levels. While expansion will occur in the long run, through changes in the deep layers due to warming, a comparatively slower impact (longer than a thousand years) will result from the melting of the icebergs; while both of these will change the level of salinity, changes in the wave regimes and the circulation in the oceans will change marine ecology. Such changes due to sea level increases will also alter the habitat. Several people will have to move inland, lower atolls and islands will be destroyed completely, and cities will need to be rebuilt. Salt water will distort underground water composition. The biological environment of the amphibians will be destroyed. The most worrying probable impact is that the carbon dioxide absorption capacity of the oceans will diminish more rapidly than the warming of the oceans, since oceans will become a source of carbon in the long run.

Furthermore, shores will be distorted, and sea traffic will move towards the north due to iceberg melt. Once sea levels have increased, algae living on the coral reefs will move upwards in order to survive by remaining at the same levels, but some will not manage this. Some people will lose their incomes. Coral reefs are also closely related to the fish population. Without coral reefs, shores will be vulnerable to the waves and sea hits. Besides this, coral reefs have touristic value.

Due to climate change, ecosystems and biological environments will be affected; while some will expand or change latitude, others will diminish or disappear. If species cannot adopt, they will migrate. As a result of distorted ecosystems and the unavailability of mass migration, some species with a lesser migration capacity will disappear. Besides this, an increase in deterioration due to warming could have a negative impact on the balance between the amounts produced and consumed. On the other hand, disparity between the life cycle of trees and the speed of global climate change will prevent a balance being reached. Iceberg melts in the Arctic, and rain instead of snow, will expose the soil which is protected from the impact of the weather, and will release methane hydrates that will accelerate the greenhouse effect. Due to the increase in the amount of carbon produced by the soil, micro-organisms and germs transforming organic materials will generate carbon emissions. Unless new forests are provided, existing forests may not be sufficient to absorb such levels of carbon dioxide through photosynthesis.

Global climate change influences agricultural production enormously. It is projected that global warming will increase inequality in the distribution of the grain. On the other hand, some regions and types will be affected positively. While productivity will increase in the Northern Hemisphere, fresh water resources and the amount of arable land in warm countries will diminish. Problems arising from floods and an increase in the frequency and intensity of storms will be observed in larger areas. Types of photosynthesis will become easier, due to the longer seasons of growth caused by global warming, and this will influence agriculture positively. Productivity of wheat in Europe may increase from 9% to 35% by 2050, which will benefit Italy, Greece, southern France and northern Spain. On the other hand, long periods of rain may destroy farms. While productivity in Portugal will decrease by 3-4 tonnes per hectare due to global warming, it will increase productivity in Northern Europe. Finland will gain 100-150 km of arable land in the north. North America will benefit from a 20-50% increase in productivity growth, and the increase will be 40-70% in Russia. However, in Africa, arable lands will diminish more rapidly, and lost areas will reach 9% by 2100, resulting in lower productivity. In Tropical Asia, each increase in the temperature risks a 7.4% decrease in production (Denhez, 2007).

Ultimately, global agricultural production will increase, but so will inequality in sharing, leading to enriched countries financing 25% more imports from poor countries. Besides this, the quality of the soil will also be affected, and changes in the aridity and degradation of soil will be observed in the same areas. New arable lands will appear in the high mountains and northern parts of the Northern Hemisphere. Hail showers will lead to refraction in products; waterspouts will cause decomposition on the branch, and an increase in parasites will reduce the amount of products. Heavy rains will take away the fertile soil. The linear relation between water and soil will increase the distance between wetlands and production areas. 5 billion people may be living in drought-affected areas by 2050. Tropical herbs, such as pearl millet, millet, and sugar cane will adjust to changing conditions (Denhez, 2007). Based on such data, the impacts of global climate change can be categorised as follows (Table 1):

**Table 1: Impacts of Global Climate Change**

| Direct Impacts (Due to Changes in the frequency and the volumes of other abnormal air conditions) | Indirect Impacts (Due to changes in the ecosystem) | | |
|---|---|---|---|
| Sudden diseases | Changes in the ecologies of water and food resources | Increase in contaminated diseases | |
| Deaths | Deprivation in water resources | Increase in contaminated diseases | |
| | Desertification | Rapid degradation of arable lands | Migration due to less agricultural areas |
| | Iceberg Meltdown and thermal expansion of sea water | Less salinisation due to sea level increase | Deterioration in the ecosystem of the shores → increase in contaminated diseases due to deprivation in water and food resources → migration |
| | Increase in virulation and reproduction the volume of disease factors | Revival of diseases (Tbc, syphilis, HIV) | |
| | Increased change in vectors' life conditions | Malaria | Arbovirus infections (Viral ensefalit, dong fever) |
| | Changes in energy resources | | |

Global climate change will deteriorate conditions on earth in terms of air quality, plant quality, food quality, water quality, natural disasters and energy resources, which all serve for survival, the main element of security; in terms of the relationship between global climate change and security, survival is primary. Through several natural disasters, nature has already announced that it is in crisis. In 1995, a series of floods led to the evacuation of 200,000 people in the Netherlands. 115 deaths and more than 150,000 injuries occurred due to high water levels in Oder in Central Europe in 1997, also resulting in evacuation. In 1998, Hurricane Mitch reflected the seriousness of the situation, with 12,000 deaths and 13,000 people lost. In China, floods led to more than 3,000 deaths. 1998 was the hottest year on record since the mid-19$^{th}$ century (when temperatures began to be archived). In 1999, one billion hectares of forests were destroyed by unexpected storms in Europe. In 2003, an increase in climate uncertainty was noted (with respect to the temperature levels in France, for example). In 2004, carbon dioxide in the atmosphere reached its highest level (379 ppm) for 700,000 years. In 2005, Hurricane Katrina destroyed northern Louisiana, causing 100 billion dollars worth of damage (Denhez, 2007). Several economists argue that these costs will increase further, and that investment is needed today. Hurricane Sandy and the cold winter in Western Europe in 2013 resulted in great losses to the economy; thus, individuals and states need to decrease their carbon emission, as reflected in the speech given by President Obama of the US (White House, 2013).

In an interview for the Milliyet Daily (2013: 16), Sandrine Dixson-Decleve (Director of Global Climate Change Leaders Group in Principality of Wales) argues that activation should work on different levels. An international framework is crucial, but developing countries tell developed countries that they do not want to stop, on the grounds that developed countries have developed excessively. In fact, China and other transition countries are aware of the serious impacts of global climate change; moreover, it is in their economic interest to invest in low carbon technology. But this is not enough. Several sectors must meet, cooperate, and share their knowledge and positive experiences in the effort to decrease carbon emissions. This requires international, regional, and local legislation.

It is estimated that unless measures are taken, the cost of global climate change will amount to 5-10 percent of Gross Domestic Product (GDP) (Stern: 2006). Thus, climate change is very closely related to economic security. In an interview for the Milliyet Daily (2013: 16), Nicolette Bartlett (Manager of Sustainable Leadership Group at Cambridge University) argues that today, controversy concerning the relation between growth and global climate change is not warranted. If the climate changes, growth is not possible. The crucial points to note here are the need to work within a market, and the pricing of carbon (the sign of the formation of a market). If the price of carbon is zero, as in Turkey, future generations will pay the price. In terms of food security, it is projected that by 2050, the global population will have increased by 2-3 billion, with a larger proportion of people on higher incomes, consuming more; it will therefore be necessary for farmers to grow twice as much as they do today (Folley, 2011: 64). When drought and flood are taken into consideration, it is argued that one billion hungry people would struggle to survive. Noam Chomsky (2013), a leading expert in politics, argues that “nobody knows the exact details, but everybody knows that the longer you wait, the worse it's going to be.”

## 4. Conclusion

Today, security is no longer solely a military issue; it is also an economic, social, political and environmental one. The Copenhagen School provides a theoretical ground and a comprehensive approach to security, which does not perceive security only in military terms, with only the state as an actor. In this theoretical framework, there exist sectors of security which are military, economic, social, political and environmental. The process of labelling an issue as a security issue is called ‘securitisation’, and is based on threat perception. Here, global climate change poses a serious threat to the main goal of human beings, survival. So, climate change is securitised, meaning that it is no longer a normal issue; instead, it is subjected to combatting policies in order to normalise it. Due to the direct relationship between threat and security, global climate change has become one of the most serious threats to security; necessary precautions are required, in order to protect the earth from the serious impacts of global climate change.

Global climate change is both an environmental and a sustainable development issue. It is mainly caused by greenhouse gases, which threaten survival in terms of air, plants, water, and food quality. Its impact has been projected over the next 20-30 years. The global commons are under threat, and global action is required in order to fight the widespread effects of climate change. Its impacts are not only on survival, for the economy is also directly affected by enormous changes to the resources necessary for production. From this perspective, environmental security and economic security are very closely related. This interdependence is most clearly observed in terms of environmental security, since this is based on the resources we need to survive, prosper, develop and grow in political, economic and social terms. Since it is related to the global commons, this responsibility should be shared globally, based on the level of damage caused by each country.

When survival is threatened, conflicts are expected to occur in regions where human beings' living conditions are affected, in terms of air quality, plant quality, food quality, water quality, natural disasters, and energy resources. Such conflicts over natural resources may reach the point of war. Recent increases in food prices are a signal of the world's instability. The world is already the scene of wars over the energy needed to control resources and energy routes. Global climate change is a challenge for the whole world, since it is based on the most basic necessity in Maslow's Hierarchy of Needs. The struggle among nations can be expected to concern the acquisition of the basic necessities for survival. What is needed in order to change this tendency is for each country to take measures to prevent deterioration in living conditions, and to support initiatives by making changes in national policy.

Due to deterioration in living conditions on earth, it is argued in the literature that this struggle will extend to space, in order to procure chemicals which are in scarce supply on earth; 'space management' is the term used to describe the obtaining of chemicals lacking on earth from space. A further recent development with regard to the rivalry between nations in ensuring the sustainability of critical resources concerns the control of the Arctic. It is projected that the Arctic contains vast amounts of valuable chemicals needed for survival, and that as a result of warming these will be released by the melting of ice (the most crucial effect of global climate change). Not only the environment, but also the economy will be affected by climate change. Due to melting, global trade routes will change. Raising awareness, generating public pressure on governments, designing targets, and pricing carbon emissions are the steps which are currently lacking, and are needed in order to combat the impacts of climate change.

## References

Buzan, B., O. Waever and J. de Wilde. (1998). Security: A New Framework for Analysis. Colorado: Lynee Riener Publishers.

Carey, R. "The Contemporary Nature of Security".

Denhez, Frèdèric (2007). Küresel Isınma Atlası. 2. Baskı. İstanbul: NTV Yayınları.

Folley, Jonathan P. (2011). "Can We Feed the World and Sustain the Planet?", Scientific American, http://www.geog.psu.edu/sites/default/files/Scientific%20American%20 Article.pdf

Houghton, R. A. (2007). "Balancing the Global Carbon Budget", Annual Review Earth Planetary Sciences, 2007, 35, pp. 313-347, http://www.geol.lu.se/events/Element/Reading_material/18.pdf

IEA, 2012 World Energy Outlook, http://www.worldenergyoutlook.org/publications/weo-2012/#d.en.26099

IPCC, Climate Change 2007: Working Group I: The Physical Science Basis, http://www.ipcc.ch/publications_and_data/ar4/wg1/en/faq-3-1.html

Milliyet Daily (29 March 2013). "Karbon Salınımı Bedava Olmamalı", p. 16.

Stern, N. (2007). The Economics of Climate Change, The Stern Review. London: Cambridge University.

White House (2013). "Remarks by the President on Climate Change", 25 June 2013 http://www.whitehouse.gov/the-press-office/2013/06/25/remarks-president-climate-change

# PARADIGM SHIFT IN MILLENNIUM DEVELOPMENT GOALS MANAGEMENT THROUGH CLIMATE CHANGE

AYLIN MET[1] AND TANER ALTUNOK[2]

## 1. Millennium Development Goals and Climate Change: Now

The MDGs (Millennium Development Goals) consist of 8 goals and 18 targets (21 in 2010). Together, these constitute an important framework for the purpose of attaining common goals of wellbeing and health—including the eradication of extreme hunger and poverty—by 2015 (United Nations). These goals were agreed upon by 191 nations at the Millennium Summit organised in September 2000 (DPT, 2010), (UNDP-1, 2011). Each of the goals is taken as an indicator for the purpose of measuring development. They have had a significant impact on the international policy agenda and are expected to continue to do so (Gonzalez, Cantero, & Dardet, 2009).

The third IPCC (Intergovernmental Panel on Climate Change) report, published in 2001, underlined that it is likely that global warming will cause changes unprecedented since the end of Ice Age. Based on knowledge gained during the 1990s, in 2001 the National Academy of Sciences committee in the US defined the sudden existence of climate change risk as a "paradigm shift" that necessitated scientific handling. Besides this, it was found that the increase in temperature measured at the bottom of the ocean matched computer simulations, and this was defined as an obvious indication of warming (AIP, 2009). The IPCC report published in 2007 showed that the impacts of climate change were widening as a result of human activity (IPCC, 2007).

[1] PhD, Scientific Programs Specialist, TÜBİTAK ÇAYDAG Atatürk Bulvarı No: 221 06100 Kavaklıdere Ankara, TURKEY, aylinseyidoglu@hotmail.com

[2] Professor of Engineering, Director, Graduate School of Natural and Applied Sciences, Çankaya University, Öğretmenler Caddesi No: 14 06530 Balgat Ankara, TURKEY, taltunok@cankaya.edu.tr

## 2. Why is Paradigm Shift Necessary?

The Millennium Summit of 2010 was organised in New York on the 20th of September, with the participation of 140 countries. It was emphasised that only 5 years remained before the deadline for the realisation of the goals set in 2000. It was also emphasised that because of the economic, food and climate change crisis the world faces today, it has become much harder to realise MDGs (OECD, 2010). It is expected that climate change will impact on the MDGs in a number of ways. These are outlined below (UNDP-1, 2011):

- Eradicate extreme poverty and hunger: Agricultural production, food security, and access to clean and adequate water resources are sensitive to climate change. Because of this, goals like creating jobs and eradicating extreme poverty and hunger are under threat.
- Achieve universal primary education: Climate change has a negative impact on livelihoods, such as agricultural production. For this reason, families may not be able send their children to school. Extreme weather may harm school infrastructure. Families may have to migrate because of limited educational facilities.
- Promote gender equality and empower women: Women engaged in hidden employment, working in their homes without salaries, and undertaking the responsibility of using and managing natural resources because of their traditional roles are the most vulnerable to climate change.
- Reduce child mortality, improve maternal health, and combat epidemic diseases: Because of the harm done to water resources by floods or drought and the change in temperatures, it will become increasingly hard to supply food and water security, which is expected to trigger epidemic diseases and threaten the health of society.
- Ensure environmental sustainability: Environmental sustainability is under threat, since climate change will cause changes in ecosystem relationships, decrease the quality and quantity of existing natural resources, and decrease the productive capacity of ecosystems. Poor nations are dependent on these resources in order to survive.
- Develop a global partnership for development: Since climate change will make the realisation of MDGs more difficult, it will be necessary to develop funding structures for development and adaptation in order to meet the needs of the poor (UNDP-1, 2011).

The picture outlined above reveals the negative impact of climate change on the successful realisation of MDGs and shows where there is as yet no adequate progress. In the management of MDGs today, where the impact of climate change is felt intensively (as compared to when the goals were first determined), it is obvious that there is a need to develop a new approach.

At this point, it will be useful to analyse the "paradigm" concept. According to Kuhn, a paradigm includes the problems and solutions within a research area over a determined time period (Kuhn, 1996). Scientifically, paradigms are used to represent concepts, models or examples formed by the filter of experience, and the common mind. Paradigms are possible with the application of the same principles, rules and standards to these concepts (Prud'homme, 2011). When the existing paradigm is not adequate for the purpose of explaining different concepts and ideas, it fails. The scientific approach which was once based on this paradigm now faces the crisis of finding a solution.

The fact that many nations came to an agreement in defining MDGs in order to increase global wellbeing can be defined as a historical breakthrough and a paradigm shift. This paradigm provided the means to determine common goals at a point when previous management approaches were not able to bring about a solution. At this point, an academic research community was formed, and many publications followed. This falls within the rules of a paradigm shift as set out by Kuhn.

However, due to climate change in particular (but also due to economic, political, social, and even cultural reasons), there have been no great achievements towards meeting the MDGs. It is now obvious that these goals will not be achieved on time. Hence, it is clear that the present paradigm is inadequate and that a new one is needed. This new paradigm, formed in response to the impact of climate change, should make creativity, growth, new ideas and new models possible, in order to produce new solutions to the problems which the old paradigm could not solve.

## 3. In Which Areas Paradigm Shift Can Be Applied?

In 2010, even though only five years remained in order to achieve the goals of the MDG, there was still a great deal left to do. But, since the world was in the process of great change, it could not be expected to stand still in the face of the achievement of these goals. The OECD gave the first signals of paradigm shift in its 2010 report, stating that "development as usual is no longer an option". It is obvious that responsibilities undertaken on the way to MDG cannot be achieved by following a cautious and narrow development agenda. Radical reforms in management approach have to include the steps presented in Figure 1-1 (OECD, 2010)

Figure 1-1 Steps of the Management Reform to Achieve MDG's

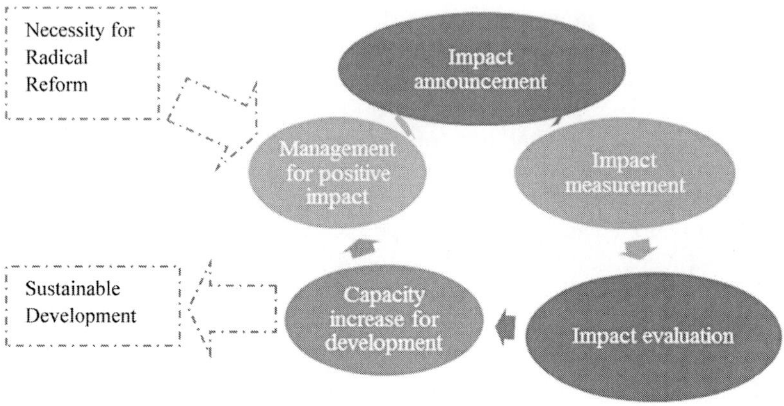


As can be seen from Figure 1-1, and the chain that begins with management for positive impact, an effective capacity increase for development can be made possible by an evaluation of applied reforms at each step. The paradigm shift in MDG management due to climate change can be analysed into its component parts, via the parameters set out in Table 1-1:

**Table 1-1 Parameters of Paradigm Shift in MDG Management by Climate Change**

| **A. Ecological Parameters:** | **B. Social Parameters:** |
|---|---|
| 1. Agricultural production | 1. Scientific development |
| 2. Econutrition | 2. Promoting multidisciplinary studies |
| **3. Saving marine ecosystem** | 3. Management applications |
| 4. Saving resources | 4. Awareness and participation |
| | 5. Infrastructure improvement |
| | 6. Improvement of the structure of donations |

## A. Ecological Parameters

**A.1. Agricultural Production:** Global climate change poses a significant threat to agricultural production and trade, and it can lead to results such as undernourishment and hunger. In many developing countries, agriculture occupies a 20-60% share of GDP (Gross Domestic Product) and 65% of employment, representing the livelihood of 2.6 billion people. In spite of the increase in food production across the world in the last decade, it does not seem possible to achieve the MDG of halving the number of people living in hunger by 2015. Models show that in the regions most vulnerable to climate change (such as sub-Saharan Africa and South Asia), a decrease in agricultural effectiveness of 15-30% (as much as 50% in some countries) is expected by the year 2080 (UNCTADPB, 2010).

UNCTAD (United Nations Conference on Trade and Development) emphasised the need for a paradigm shift in agricultural production, in order for the agricultural sector to become part of the solution by decreasing the impacts of climate change and adaptation in developing countries (CCPP, 2010). The potential for climate change to do irrevocable harm to natural resources was underlined in the policy summary. Hence, a change in the agricultural production paradigm has importance for food security. In this way, the agricultural sector could cease to be a problem in the climate change challenge and become part of the solution. *Shifting from traditional industrial agriculture dependent on one culture to an approach which depends upon providing facilities to small scale farmers in order to increase their production capacity is important.* But while a sudden change to the existing agricultural system cannot be expected, developed countries can bring about targeted agricultural and fiscal policy changes and strengthen their sustainable agricultural policies through promotion mechanisms (CCPP, 2010).

Development of agricultural policy is a process which necessitates the participation of experienced multidisciplinary experts. Hence, the action plans developed in order to support agriculture should have a national and international framework. *In order to develop a sustainable agriculture, evaluations of agricultural scientific and technological areas should be carried out, and suggestions should be developed.* National and international action plans related to sustainable agriculture are presented in Figure 1-2 (UNCTADPB, 2010):

Figure 1-2 National and International Level Action Plans for Sustainable Agriculture

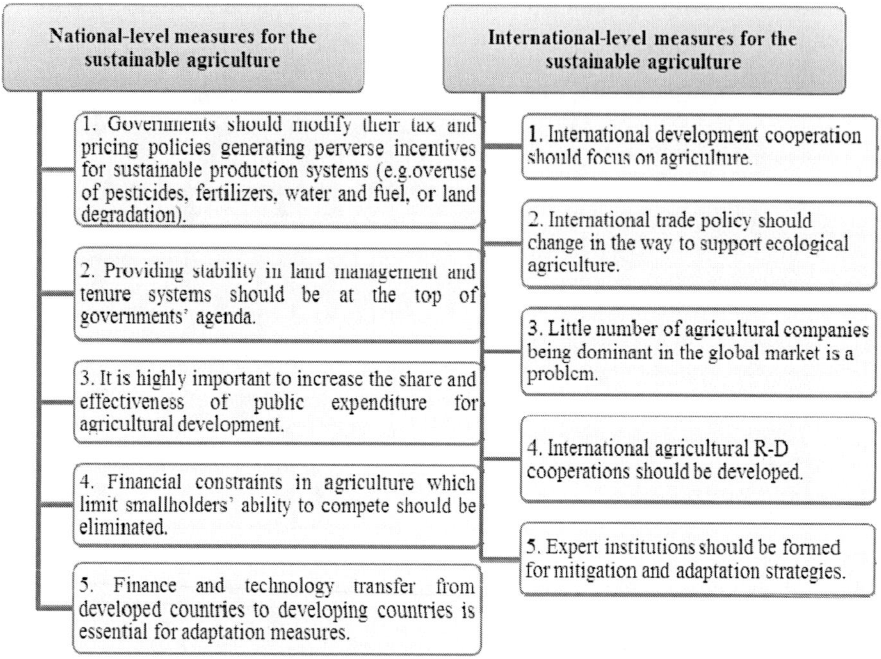


**A.2. Econutrition:** Econutrition is a concept that integrates environmental health and human health, focusing on the interaction between agriculture, ecology and nutrition. Malnutrition caused by the decrease in the quality and quantity of soil is one of the most important barriers facing economic development. Econutrition is an example that shows why multidisciplinary studies should be carried out. Research performed in this area includes not only ending poverty, but also fundamental science. *In order to gain, at low cost, in the area of human health and agricultural efficiency, multidisciplinary and multisectoral interventions from a related thematically perspective should be carried out.* In 2015, despite a decrease in malnutrition in Southeast Asia, East Asia and Latin America, the story will be very different in sub-Saharan Africa (Deckelbaum, Palm, Mutuo, & DeClerck, 2006). For this reason, it should be possible to reduce the number of societies facing hunger by implementing the results of studies on econutrition.

**A.3. Saving Marine Ecosystem:** UNDP (United Nations Development Programme) agrees that a paradigm shift in marine ecosystem management needs to occur, with the aim of changing the climate; further achievable goals should also be pursued, such as eradicating poverty, ending hunger, and providing a sustainable use of the environment. The marine ecosystem has a 12.6 trillion dollar share in the world economy, and marine ecosystem products provide a livelihood for more than 1 billion people around the world, mostly the poorest people living in coastal areas of Africa, Asia, Latin America, and Eastern Europe. Countries should develop support mechanisms in order to create the sectorial capacity to allocate resources evenly (UNDP-2, 2010).

**A.4. Saving Resources:** Ensuring the goal of environmental sustainability has a critical impact on the achievement of other goals. The livelihood of many poor communities depends on natural resources. Capital founded on natural resources generates 26% of poor countries' income. On the other hand, 20% of diseases in developing countries are caused by low environmental quality. From this perspective, it is possible to say that natural resources are important factors in alleviating poverty. Ensuring environmental sustainability is possible with the integration of sustainable development principles within the policies, programmes and strategic plans of a country (UNDP-3, 2011). Since this approach is emphasised in much of the relevant research, sustainability seems to be a bottom-up approach. However, the important thing is the integration of these principles by each individual member of the community into their daily lives. *Saving resources is only possible with the application of bottom-up management principles. For this, community awareness should be increased.* A basic rule for the achievement of government policies is that these policies should be adopted and voluntarily applied at every level of the community. Increasing community awareness is a government responsibility.

## B. Social Parameters

**B.1. Scientific Development:** Each member of the UN declared their commitment to the achievement of MDGs in 2015. This means new research and the implementation of new scientific developments. Among world leaders, it is agreed that *science is an important component of development*. Acharya et al., (2003) proved that there is a strong relationship between scientific invention and development (Acharya, Daar & Singer, 2003). For example: is there a relationship between progress in biotechnology and MDG? In seeking an answer to this question (where the relationship between progress in biotechnology and MDG is analysed with the help of parameters like the impact of technology, appropriateness for developing countries, the burden to the country of the search for solutions to the problem, as well as benefits like gaining new knowledge), it was found that the existence of the relationship is obvious. However, the most important barrier facing this relationship is the distribution of the most appropriate technology needed for the purposes of development across a large area. For example, when we deal with the goal of "ensuring environmental sustainability," progress in biotechnology will be reflected as "bioremediation" in the environmental area. One benefit of bioremediation is that microbes or plants are used to determine environmental pollutants, and to degrade them at low cost and in a sustainable way. For example, NT-26, which transforms arsenate into less soluble and less toxic arsenate, may help to clear arsenic-contaminated water (Acharya, Daar&Singer, 2003). From this perspective, encouraging scientific research in areas like biotechnology and creating the necessary infrastructure will help in the achievement of MDGs.

**B.2. Promoting Multidisciplinary Studies:** *Multidisciplinary studies bringing together many sectors in order to improve the health, nutrition, agriculture, water, education, transport and environmental sustainability should be promoted.* The Millennium Villages Project of Earth Institute at Columbia University is a good example of this (Deckelbaum, Palm, Mutuo, & DeClerck, 2006). The primary aim of the project is to bring together many sectors and to achieve the eradication of extreme poverty and hunger in those regions of Africa that face difficulties. Villagers benefitted from this multisectorial enterprise, and scientific progress was made. For example, research was done into agricultural production, water cycles and soil productivity, and knowledge of ecosystem processes was gained. Through the application of a multidisciplinary structure, it was found that (for example) public health and political epidemiology could be used in order to increase knowledge in the area of health (Gonzalez, et al., 2006). *The shift from single-discipline studies to a multidisciplinary approach can be taken to be a paradigm shift.*

**B.3. Management Applications:** The UN is in the phase of examining why the MDGs have not been achieved. It is thought that political and social factors have played a role in this situation. Existing public policy should be evaluated and stronger targets should be determined, analysing the subject from every perspective. *Adding a political and social point of* *view into MDG is an indicator of paradigm shift* (Gonzalez, Cantero, & Dardet, 2009). Considering the watershed management example (in areas like the Mekong River, which includes riparian countries), the traditional watershed management approach considers water as a mere resource. But this approach should be transformed into one that considers water as a resource which creates resources (Rothe, 2010).

Shifting to a complicated adaptation system may introduce some handicaps, because existing management systems are quite rigid. If it can be done, triggering the change will enable hidden competencies to emerge (Hunter, 2009). For this reason, the management paradigm has to change, via two steps. The first is to provide a proper medium for allowing paradigm shift to happen. The second is to provide for the realisation of better adaptation applications, using the results of this shift. Management applications require three interrelated steps, because of their case dependent nature: (i) improving the system, (ii) improving the leadership definition for management, and (iii) improvement of knowledge and competencies. Policy analyses should be carried out at each step, and their organisational structure should be evaluated adequately (Hunter, 2009).

**B.4. Awareness and Participation:** Besides the change in management paradigm, it is also necessary to "*create awareness and provide participation in the society*". Society bears a responsibility for achieving MDGs, and it is society that will be most affected if the MDGs are not achieved. Society should be widely informed about the methods used to achieve MDGs and why such methods were chosen (Kassim & Ali, 2006). Not being able to provide awareness in society is the most important factor in failure. *Private sector contribution is as important as public sector contribution in increasing society's awareness of the achievement of MDGs*. Missions and responsibilities should be well defined within the public sector, the private sector, and society. Even though there is an achievement potential, barriers facing this achievement should be investigated.

**B.5. Infrastructure Improvement:** In economies where living difficulties arise due to the negative effects of climate and geography, public enterprise and infrastructure investment are needed, in order to handle these difficulties. Through an analysis of the effects of infrastructure on growth, it can be seen that infrastructure improvement has an important role in promoting development (Ndulu, 2006). Through infrastructure improvement, the poorest parts of a society are able to access electricity, clean water, and roads connecting urban and rural areas. It was a historical breakthrough that a focus on the need to improve infrastructure (i.e., wider social services and the elimination of challenges posed by scarce state resources) was put forward via the MDGs. *A major structural change in public infrastructure investments is that regulatory bodies apply participatory management principles together with the users of services, and the participation of the private sector in public sector investments becomes possible* (Ndulu, 2006).

**B.6. Improvement of the Structure of Donations:** The Development Co-operation Report, published by the OECD (2010), underlines that the achievement of MDGs became much more difficult in the light of the recent economic, food and climate change crises (OECD, 2010). The report emphasised that the "*development dimension of the crises should be a part of the political agenda*". Using development aid (especially during periods of economic uncertainty) to strengthen capacity has great importance. This aid has to be conveyed to the poorest regions and should not be conditional. It is understood that member countries have to develop a broader and more comprehensive approach, and include climate change in their political decision-making. The positive impact of global factors other than aid on development emphasises the importance of reform and adaptation efforts that go beyond the traditional concept of aid.

In order to provide sustainable environmental production, national and international action plans should be developed (UNCTADPB, 2010). Because the implementation of MDGs has slowed down due to climate change, paradigm shifts such as those emphasised above should be globally implemented.

Another important point concerns the job creation effect of the 2005 Kyoto Protocol, with the help of all these components. Resilient mechanisms of the Protocol aim to decrease global greenhouse gas emissions and to trigger global capital movements and technology transfer. Consequently, it is predicted that in many areas (primarily the area of renewable energy), new employment opportunities can be created (UNDP-4, 2009). *New approaches developed for the achievement of MDGs will contribute to employment creation, which will assist economic improvement*. This target depends on the strong political determination of all shareholders, who need to make the necessary political changes and take concrete steps forward. Within this framework, taking concrete steps is necessary in order to increase the possession of developing countries, to provide the predictability of development funding, and to develop innovative funding sources and effective uses of development sources (DTM, 2010).

In Turkey, environmental strategies have been included in the Five Year Development Plans that have been developed since 1991. The ninth Five Year Development Plan, (2007-2013) which is currently effective, brings the need to prepare a National Climate Change Action Plan to the agenda (Atamer, 2010). Turkey, located in the Mediterranean Basin, is among the countries that are highly vulnerable to climate change, according to UNFCCC (United Nations Framework Convention on Climate Change) predictions (UNDP-4, 2009). Turkey's Millennium Development Targets Report was prepared by DPT (State Planning Organisation) in 2010, with the technical support of the UNDP and the UN. As in the first report, published in 2005 (İstanbul MDG, 2010), it was seen that Turkey is in many respects on the way towards achieving MDGs (DPT, 2010). However, the report also emphasised that some inequalities that are dependent upon social gender and geographical issues are the difficulties most likely to arise.

## 4. Conclusion and Discussion

The MDG is a plan to increase global wellbeing. The definition of action focus of the MDG is itself a sign of a historical breakthrough (Gonzalez, et al., 2006). On the other hand, revealing the existence of climate change has triggered a new paradigm. The impact of climate change (when connected to economic, social and political components) has been shown to have a negative effect on the achievement of MDGs. Despite the benefits of these internationally agreed goals, they have yet to be fully achieved. Hence, due to the impact of climate change, it is obvious that there is a need for conceptual and methodical change in the management approach of MDG: this means that a paradigm shift is required. In the present study, the paradigm shift in MDG management through climate change was analysed under two components (ecological parameters and social parameters). The findings related to each parameter are presented in Figure 1-3 (adapted from UNCTADPB (2010). Each component of the system creates an interrelated relationship web. This shows, from the stabilisation of the system point of view, that focusing on only one or some of the components is not enough.

Figure 1-3 Paradigm Shift Component Web in MDG Management

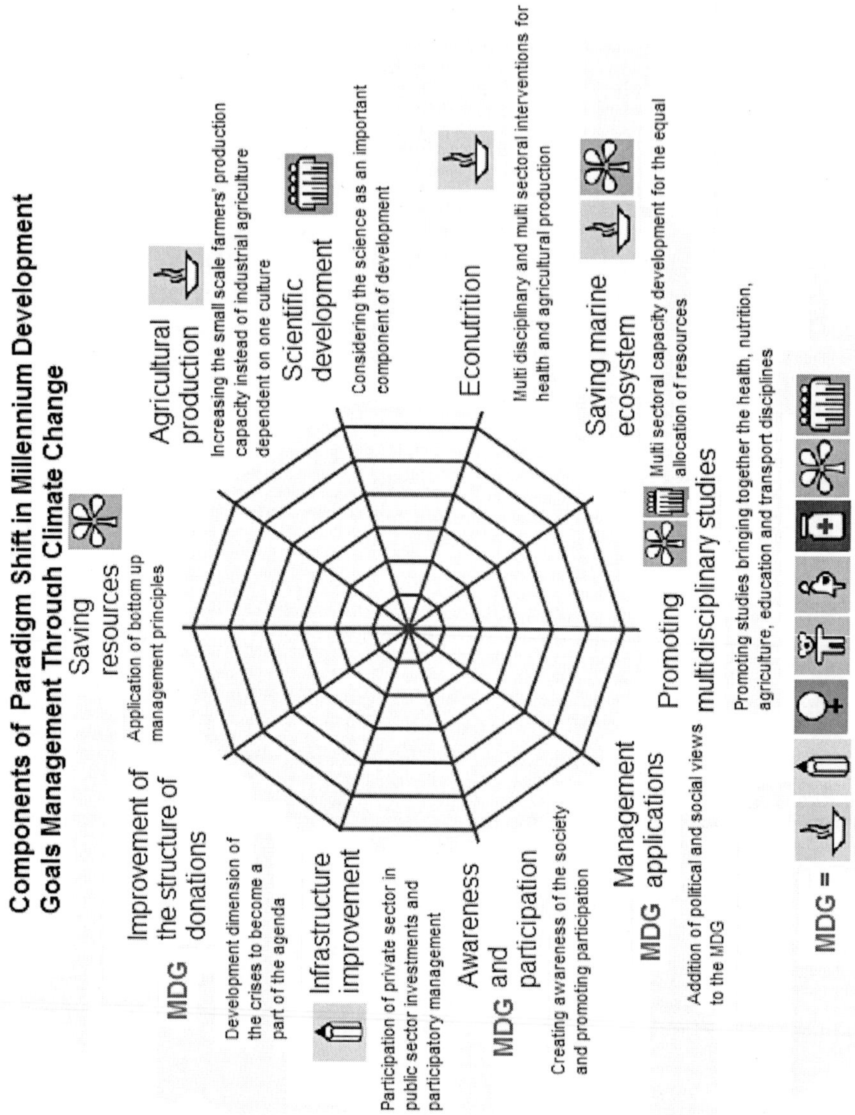

As can be seen in Figure 1-3, the impact of economic crises on development should be analysed by the addition of political and social points of view to the MDGs. Increasing small scale farmers' production capacity (instead of industrial agriculture dependent on one culture in agricultural production) is important. Providing an equal allocation of resources is important, if the marine ecosystem is to be saved. Promoting studies bringing together health, nutrition, agriculture, water, education and transport disciplines is essential for the development of multidisciplinary and multisectorial support mechanisms. Awareness should be established in society and bottom up management principles should be promoted, in order to provide participation and save resources. The responsibilities of the public sector should be balanced with the participation of the private sector. Within this paradigm shift, the most important component is the consideration of science as a basic principle of development. By considerating each principle as interrelated with others in a relationship web, making progress towards these goals becomes possible.

In all precautions against climate change, *society based adaptation attempts form an important component of national adaptation efforts*. Without the participation of local actors, legal regulations will be useless. Hence, a framework for global cooperation should be established, necessitating local action for sustainable development and environment (Vandeweerd, 2009). As a matter of fact, during the shift to a new paradigm, process concepts (like the mitigation of the effects of climate change and adaptation to these effects) came to the MDG agenda, and the importance of scientific and technological developments was emphasised (Rothe, 2010). The inadequacy of one-discipline studies in proposing solutions opened the way to the development of support mechanisms for multidisciplinary and interdisciplinary studies. On the other hand, the most important component of MDG (environmentally sustainable development) is not a simple goal. Since the day it was defined, the most strongly emphasised part of sustainable development concerns the *cooperation of society, government, university, and industry* (Rothe, 2010).

Research has shown that unless goals are adopted by society, they cannot be achieved. This is the reason why *the determination of society must be integrated into the studies*. Providing the awareness and participation of society is the most essential step in the development of local capacity. For the development of local capacity, principles of mitigation and adaptation for climate change and resource properties should be determined, and actions should be defined primarily at the regional, national, and local levels. Funding mechanisms to support these actions should be developed, and these steps should be integrated into

Figure 1-4 Steps for the Improvement of Local Action Capacity

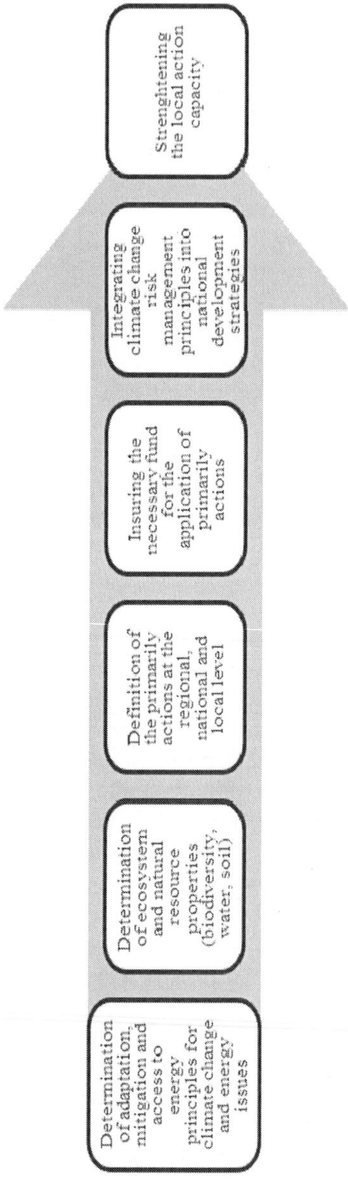

national climate change strategic plans. Achieving effective results depends on the rule of creating cooperation among different sectors. Inviting institutions to participate in the realization of these goals is essential to their development (Gonzalez, et al., 2006). In this way, local action capacity will be strengthened, and an important step will be taken towards the achievement of the MDG. Steps for the improvement of local action capacity are given in Figure 1-4 (Vandeweerd, 2009):

Combating climate change is a process which necessitates the integration of the issue into national strategies and plans, and the creation of policies to finance solutions (UNDP-5, 2011). Not being able to completely achieve the MDGs because of climate change led the UNDP to place climate change at the centre of their policy targets, and to carry out a series of capacity development attempts. Through these attempts, combat and adaptation strategies necessitating a national transformation approach were improved in developing countries and vulnerable regions of the world (UNDP-1, 2011).

## References

Acharya, T., Daar, A. S., & Singer, P. A. (2003). Biotechnology and the UN's Millennium Development Goals. Nature Biotechnology, 21 (12), pp. 1434-1436.

AIP. (2009, July). Timeline (Milestones). Retrieved January 19, 2011, from http://www.aip.org/history/climate/timeline.htm

Atamer, S. A. (2010, August). İklim Değişikliği Politikaları Mevcut Durum Değerlendirmesi Raporu (2. taslak). Retrieved February 01, 2011, from http://www.iklim.cob.gov.tr/iklim/Files/Iklim%20Degisikligi%20Politikalari%20Mevcut%20Durum%20Degerlendirmesi%20Raporu.pdf

CCPP. (2010, December 3). UNCTAD Recommends Paradigm Shift to Make Agriculture Climate-Friendly. Retrieved January 19, 2011, from http://climate-l.iisd.org/news/unctad-recommends-paradigm-shift-to-make-agriculture-climate-friendly/

Deckelbaum, R. J., Palm, C., Mutuo, P., & DeClerck, F. (2006). Econutrition: Implementation Models from the Millenium Villages Project in Africa. Food and Nutrition Bulletin, 27 (4), pp. 335-342.

DPT. (2010). Binyıl Kalkınma Hedefleri Raporu Türkiye 2010. Ankara.

DTM. (2010, June 17). AB Devlet ve Hükümet Başkanları Zirvesine İlişkin Not. Retrieved January 21, 2011, from www.dtm.gov.tr/dtmadmin/upload/AB/SanayiSektorDb/17hazzirve.doc

Gonzalez, D. G., Cantero, M. R., & Dardet, C. A. (2009). How political epidemiology research can address why the millennium development goals have not been achieved: developing a research agenda. J Epidemiol Community Health, 63, pp. 278-280.

Gonzalez, D. G., Solis, M. P., Cantero, M. T., Moncada, M. D., Giraldo, A. F., Stein, A., et al. (2006). The Challenge to Public Health of the Millenium Development Goals: An Approach from Political Epidemiology. Gaceta Sanitaria, 20 (2), pp. 61-65.

Hunter, D. J. (2009). Leading for Health and Wellbeing: The Need for a New Paradigm. Journal of Public Health, pp. 1-3.

IPCC. (2007). IPCC Fourth Assessment Report: Climate Change 2007 (AR4) 20 March 2011. Retrieved April 17, 2011, from IPCC web site: http://www.ipcc.ch/publications_and_data/publications_and_data_reports.shtml#1

İstanbulMDG. (2010, May 11). Binyıl Kalkınma Hedefleri +10 Bağımsız Devletler Topluluğu Bölgesel Konferansı. Retrieved February 08, 2011, from http://tr.istanbulmdg2010.org/binyil-kalkinma-hedefleri/turkiyedeki-aktiviteler/

Kassim, S. M., & Ali, M. (2006). Solid Waste Collection by the Private Sector: Households' Perspective-Findings from a Study in Dar es Salaam City, Tanzania. Habitat International, 30, pp. 769-780.

Kuhn, T. S. (1996). The Structure of Scientific Revolutions (3. Press). Chicago ve Londra: The University of Chicago.

Ndulu, B. J. (2006). Infrastructure, Regional Integration and Growth in Sub-Saharan Africa: Dealing with the Disadvantages of Geography and Sovereign Fragmentation. Journal of African Economies, 15 (2), pp. 212-244.

OECD. (2010). Development Co-operation Report 2010.

Prud'homme, A. (2011). Paradigm Shifts and Science. Retrieved January 19, 2011, from http://www.articlesnatch.com/Article/Paradigm-Shifts-And-Science/905204

Rothe, A. (2010). Need for a paradigm shift in Mekong management. Retrieved January 19, 2011, from http://www.eastasiaforum.org/2010/07/15/need-for-a-paradigm-shift-in-mekong-management/

UNCTADPB. (2010, December 18). UNCTAP Policy Briefs, Agriculture at the Crossroads: Guaranteeing Food Security in a Changing Global Climate. Retrieved January 20, 2011, from http://www.unctad.org/en/docs/presspb20108_en.pdf

UNDP-1. (2011). Climate Change and the Millennium Development Goals. Retrieved January 18, 2011, from http://www.undp.org/climatechange/cc_mdgs.shtml

UNDP-2. (2010, November 17). Sustainable Development of the World's Large Marine Ecosystems. Retrieved February 03, 2011, from http://www.undp.org/water/

UNDP-3. (2011, January 17). Çevre ve Sürdürülebilir Kalkınma. Retrieved January 17, 2011, from http://www.undp.org.tr/Gozlem3.aspx?WebSayfaNo=325

UNDP-4. (2009, June). Türkiye'nin İklim Değişikliği Ulusal Eylem Planının Geliştirilmesi Projesi (Haziran 2009-Eylül 2010). Retrieved February 01, 2011, from http://www.undp.org.tr/Gozlem3.aspx?WebSayfaNo=2058

UNDP-5. (2011). Integrating climate change into development. Retrieved January 18, 2011, from http://www.undp.org/climatechange/integrating_cc.shtml

Vandeweerd, V. (2009). Local Solutions for Poverty, Environment, Climate Change, and the MDGs: UNDP's Approach to Building and Catalyzing Local Capacity and Action. Retrieved February 03, 2011, from 14th Poverty Environment Partnership (PEP) Meeting: http://www.slideshare.net/PovertyEnvironmentNet/local-solutions-for-poverty-environment-climate-change-and-the-mdgs-undps-approach-to-building-and-catalyzing-local-capacity-and-action

# SEEKING SECURITY IN A CHANGING CLIMATE

ARGIRO VASILAKI[1]

## 1. Climate Change and Security

Vulnerability to the specific impacts of climate change is at its most severe when and when those impacts are felt together with stresses from other sources. Non-climatic stresses may include poverty, unequal access to resources, food security, environmental degradation, and risks from natural hazards (IPCC, 2007).

Security is affected both by changes in provisioning services—which affect supplies of food and other goods, and may increase the likelihood of conflict over declining resources—and by changes in regulating services, which could influence the frequency and magnitude of floods, droughts, landslides, or other catastrophes (MEA, 2006).

Climate change and its adverse effects on the economy and society pose significant threats to security. Whether it is a matter of food security, energy security, human security, or even national security, climate change combined with other relevant characteristics will disrupt the delicate balance of stability. Unpredictable risks are likely to emerge from a combination of demographic pressure, climate change, and resource scarcity (WDR, 2011). The avoidance of dangerous climate change has therefore become one of the biggest challenges facing humankind (WBGU, 2011).

According to the official documents of the European Union and the United Nations, climate change is best viewed as a threat multiplier which exacerbates existing trends, tensions, and instability.

[1] Economist, PhD Student, University of the Aegean, GREECE, avasilaki@aegean.gr

## 2. Official UN Documents—Climate Change and its Security Implications

Climate change was first raised as an issue in the UN Security Council on 17 April 2007. This was the Security Council's first ever open debate exploring the relationship between energy, security, and climate. The discussion focused on the security implications of climate changes, including their impact on potential drivers of conflict, such as access to energy, water, food, and other scarce resources, as well as population movement and border disputes. Secretary-General Ban Ki-moon stated that projected climatic changes might not only have serious environmental, social, and economic implications, but implications for peace and security as well (SC/9000).

In its sixty-third session, in June 2009, the General Assembly of the United Nations expressed deep concerns about the possible security implications of climate change (A/63/PV.85) and in September 2009, during its sixty-fourth session, it identified five channels through which climate change could affect security. These are presented in the report *Climate Change and its Possible Security Implications* (A/64/350). The first channel links climate change impacts to threats to the well-being of the most vulnerable communities. The second channel links climate change impacts to economic development. The third refers to the second-order effects of unsuccessful adaptation, in the form of the uncoordinated coping or survival strategies adopted by local populations. The fourth highlights the threat posed by climate change to the viability and even the survival of a number of sovereign states, in particular through sea-level rises and the resultant loss of national territory. The last channel connects climate change to changes in natural resource availability or access and the resultant competition between countries, which may include territorial disputes.

**(a) Vulnerability:** Climate change threatens food security and human health, and increases human exposure to extreme events;
**(b) Development:** If climate change results in the slowing down or the reversing of the development process, this will exacerbate vulnerability and could undermine the capacity of states to maintain stability;
**(c) Coping and security:** Migration, competition over natural resources, and other coping responses of households and communities faced with climate-related threats could increase the risk of domestic conflict, as well as having international repercussions;

**(d) Statelessness:** There are implications for rights, security, sovereignty, and the loss of statehood due to the disappearance of territory;
**(e) International conflict:** There may be implications for international cooperation from the impact of climate change on shared or undemarcated international resources.

At its 6479th meeting, on 11 February 2011, the Security Council stressed the need to consider economic, social, and political factors in maintaining international peace and security. The effects of climate change were addressed as factors that have increasingly serious impacts on development and security (SC/10172).

## 3. European Union: Climate Change is a Threat Multiplier

The European Security Strategy document, *A Secure Europe in a Better World* (2003: 3), acknowledged that "Competition for natural resources—notably water—which will be aggravated by global warming over the next decades, is likely to create further turbulence and migratory movements in various regions".

The European Council has drawn attention to the impact of climate change on international security. In June 2007, the council invited High Representative Javier Solana and the European Commission to author a joint report on the issue. This report, entitled *Climate Change and International Security*, was presented in March 2008. It analysed the consequences for international security resulting from the adverse effects of climate change (S113/08). The report outlines some of the forms of conflicts driven by climate change that may occur in different regions of the world: conflict over resources, economic damage, risk to coastal cities, damage to critical infrastructure, loss of territory, border disputes, environmentally induced migration, situations of fragility and radicalisation, tensions over energy supply, and pressure on international governance. Climate change is here viewed as a threat multiplier which exacerbates existing trends, tensions, and instabilities.

The report on the implementation of the European Security Strategy, *Providing Security in a Changing World* (S407/08), identified a range of threats and challenges to EU security interests. Climate change and energy security are among the five most serious threats:

> Natural disasters, environmental degradation and competition for resources exacerbate conflict in situations of poverty and population growth, with humanitarian, health, political and security consequences, including greater migration. Climate change can also lead to disputes over trade routes, maritime zones and resources previously inaccessible (p. 5).

In February 2010, the Council of the European Union presented *Towards a European Security Model*, a draft Internal Security Strategy for the European Union, and stated that "natural and man-made disasters, such as forest fires, earthquakes, floods and storms, droughts, energy shortages and major Information and Communication Technologies (ICT) breakdowns, pose safety and security challenges."

## 4. Climate Change and Food Security Nexus

Direct effects of climate change will be caused by changing patterns of precipitation, snow melt, and rising sea levels, which will affect hydrology and water quality. Indirect effects will result from changing vegetation patterns that may alter the food chain and increase soil erosion (IPCC, 2007).

According to the Food and Agriculture Organisation of the United Nations, "food security exists when all people, at all times, have physical, social and economic access to sufficient, safe and nutritious food which meets their dietary needs and food preferences for an active and healthy life" (FAO, 2003).

Climate change will affect all four dimensions of food security: availability, accessibility, stability, and utilisation. The adverse effects of climate change will have a direct impact on the output of production. The main components of the production process are labour, capital, and land: "all three will be affected by climate change—be it through the damaging effects on the health and productivity of the labour force, the loss and damage to agriculture and infrastructure, or lower quality investment and capital" (Stern 2007). The negative effects of climate change on land equate to the declining quality of the environment. Deteriorating environmental quality affects the output of production, with consequences for food security and the employment base of many people. Extreme weather conditions, rising temperatures, and different patterns of precipitation will affect the fertility of soils. Rising sea levels, desertification, and urbanisation will eventually contribute to the loss of land, with economic and social consequences.

Climate change will thus act as a multiplier of existing threats to food security: it will make natural disasters more frequent and intense, land and water more scarce and difficult to access, and increases in productivity even harder to achieve (IASC, 2009). In 2010, for example, Russia experienced its hottest summer since records began, 130 years ago. Droughts and wildfires covered an area of one million hectares, killed thousands of people, destroyed thirty per cent of the country's wheat crops, and caused damage amounting to a cost of 15 billion dollars. Russia, formerly the world's third biggest grain exporter, sent wheat prices soaring when it introduced a temporary ban on these exports on 15 August 2010. The embargo was introduced in order to ensure domestic supply (FAO, 2010).

Among the factors that triggered social unrest in the Middle East and North Africa in 2011 were high food prices combined with poverty, high rates of unemployment, and low incomes.

According to the FAO, global food prices increased for the eighth consecutive month in February 2011, reaching the highest level ever recorded. The FAO Food Price Index averaged 236 points in February 2011, and 230 points in March 2011, having increased 155 per cent from the year 2000 (see Table 1-2). Global food prices are now at dangerously high levels (Zoellick, 2011).

**Table 1-2 Food Price Index 2000-2011**

| Date | Food Price Index | Meat Price Index | Dairy Price Index | Cereals Price Index | Oils Price Index | Sugar Price Index |
|---|---|---|---|---|---|---|
| 2000 | 90 | 96 | 95 | 85 | 68 | 116 |
| 2001 | 93 | 96 | 107 | 86 | 68 | 123 |
| 2002 | 90 | 90 | 82 | 95 | 87 | 98 |
| 2003 | 98 | 97 | 95 | 98 | 101 | 101 |
| 2004 | 112 | 114 | 123 | 107 | 112 | 102 |
| 2005 | 117 | 120 | 135 | 103 | 104 | 140 |
| 2006 | 127 | 119 | 128 | 121 | 112 | 210 |
| 2007 | 159 | 125 | 212 | 167 | 169 | 143 |
| 2008 | 200 | 153 | 220 | 238 | 225 | 182 |
| 2009 | 157 | 133 | 142 | 174 | 150 | 257 |
| 2010 | 185 | 152 | 200 | 183 | 193 | 302 |
| 1/2011 | 231 | 166 | 221 | 245 | 278 | 420 |
| 2/2011 | 236 | 170 | 230 | 254 | 279 | 418 |
| 3/2011 | 230 | 169 | 235 | 252 | 260 | 372 |

**Source:** FAO

## 5. Environmentally Displaced People and Social Unrest

Natural disasters, extreme weather conditions, desertification, and rising sea levels are having serious impacts on the livelihoods of people, with concomitant economic and social effects. In Alexandria, Egypt, for example, the scenario of a 0.3 meter rise in sea level would lead to the loss of employment for 70,465 people, would displace 545,000 people, and contribute to a total loss of 19 $km^2$ of land (Agrawala et al., 2004).

Population growth, urbanisation, climate change, water scarcity, and food and energy insecurity are exacerbating conflict and combining in various ways that oblige people to flee their countries of residence (Guterres 2010).

Due to the lack of an agreed definition in international law or a UN definition, this report adopts the definition given by Brian Gorlick, a UNHCR Senior Policy Advisor. According to Gorlick, environmentally displaced persons are "people who are displaced from or who feel obliged to leave their usual place of residence, because their lives, livelihoods and welfare have been placed at serious risk as a result of adverse environmental, ecological or climatic processes and events" (Gorlick 2007).

According to the Inter-Agency Standing Committee, different types of displacement derive from various environmental causes (IASC 2008: 2):

> Hydro-meteorological extreme hazard events lead to either temporary forced displacement within national or across international borders; temporary voluntary movement across international borders; or forced displacement as a result of areas being designated as prohibited for habitation by authorities, potentially resulting in internal displacement, forced cross-border movements and/or voluntary cross-border movements. Environmental degradation and/or slow-onset extreme hazard events will likely be gradual, beginning with voluntary movements (within and outside the country) and potentially ending in forced displacement (within and outside the country). Environmental degradation, whether at early or advanced stages, and/or slow-onset disasters may also result in areas being prohibited for habitation by authorities, leading to internal displacement, forced cross-border movements, and/or voluntary cross-border movements. Significant permanent losses in state territory (sea level rise), if not prevented by sufficient mitigation, could be gradual, beginning with voluntary movements (within and outside the country) and potentially ending in forced displacement (within and outside the country). These could include: voluntary movements in-country (to safe parts of the country) and across internationally recognised borders; displacement within national territory; and forced cross-border movements, including in extreme cases the entire loss of state territory.

> Armed conflict/violence over shrinking natural resources can lead to forced internal displacement or to people crossing international borders as refugees or persons under temporary or subsidiary forms of protection.

People in all the above situations will have to make difficult decisions about where and how they will build up their new livelihoods. Whether they move within or outside national territory, social unrest is likely to occur. Social unrest will occur due to disappointment in the ways in which their political leaders deal with the situation or mitigate the adverse effects of climate change, or may be due to social upheaval in the receiving areas resulting from the scarcity of resources and competition between the displaced people and local communities. Grievances and resource competition, combined with a lack of representative institutions, economic redistribution mechanisms, and poor state capacity to deter violence, present the greatest risk of conflict (Gleditsch et al., 2007). Although environmental factors and scarcity of land resources have shaped the living conditions of people throughout history, and migration has long been a strategy of people facing climate change, the present conditions differ from those faced in the past. The effects of climate change are different, natural resources are becoming scarcer, and countries have strict policies for receiving migrants and refugees.

Taking into account the fact that developing countries are most exposed to the adverse effects of climate change—even though they have contributed the least to greenhouse gas emissions—as well as the present unrest in North Africa and the Middle East, climate change will intensify existing problems. The poorest of people are those who depend most on natural resources for their livelihoods. Due to their low incomes, if they are urged to seek land, they will probably move within the country to communities already facing environmental scarcity. Competition and unrest over resources is probably inevitable.

## 6. Conclusion

Climate change is an issue of international political importance. The Brundtland Commission Report, which seems to be more topical than ever, pointed out that:

> Environmental threats to security are now beginning to emerge on a global scale. Any such climatic change would quite probably be unequal in its effects, disrupting agricultural systems in areas that provide a large proportion of the world's cereal harvests and perhaps triggering mass population movements in areas where hunger is already endemic. Slowing,

or adapting to, global warming is becoming an essential task to reduce the risks of conflict (A/42/427).

Climate change is viewed by international organisations as a threat multiplier, exacerbating existing trends and tensions and creating new ones. Many nations are already experiencing environmental insecurity. They fail to effectively govern, manage, and utilise their natural resources and environment, resulting in social, economic or political instability that over time may lead to heightened tensions, social turmoil, and conflict (Deligiannis, 2010). People's livelihoods are under threat from desertification, drought, natural disasters, sea-level rises, and water scarcity. Environmental scarcity has insidious and cumulative social impacts, such as population movement, economic decline, and the weakening of states (Homer-Dixon, 1994).

Climate change is a security issue without military solutions. Today's security environment is volatile and unpredictable, and climate change is often a silent threat—it does not appear until its consequences are irreversible. The intensity and complexity of climate-related problems requires coordinated action from the international community, and the strong political will of nation-states.

## References

### Primary Sources

Council of the European Union. 2010. *Towards a European Security Model*. Draft internal security strategy for the European Union (58/42/2/10 REV2). Brussels, 23 February 2010.

European Council. 2003. *A Secure Europe in a Better World*—European Security Strategy. Brussels, 12 December 2003.

—. 2008. *Providing Security in a Changing World*—Report on the implementation of the European Security Strategy (S407/08). Brussels, 11 December2008.

European Commission. 2008. *Climate Change and International Security*. Paper from the High Representative and the European Commission to the European Council (S113/08). Brussels, 14 March 2008.

Guterres, Antonio. 2010. Speech of the High Commissioner for Refugees. Dialogue on protection gaps and responses. Palais des Nations, Geneva, 8 December 2010.

UN General Assembly. 1987. *Our Common Future*. Report of the World Commission on Environment and Development, forty second session (A/42/427). 4 August 1987.

—. 2007. Letter dated 5 April from the Permanent Representative of the United Kingdom of Great Britain and Northern Ireland to the United Nations addressed to the President of the Security Council (S/2007/186). 5 April 2007.

—. 2007. 5663rd meeting (S/PV.5663), Provisional. New York, 17 April 2007.

UN General Assembly. 2009. Official records, 85th plenary meeting, 63rd session (A/63/PV.85). New York, 3 June 2009.

—.2009. Climate change and its possible security implications (A/64/350). New York, 11 September 2009.

UN Security Council. 2006-7. Report of the UN Security Council, 1 August 2006-31 July 2007 (A/62/2), General Assembly Official Records, 62nd Session Supplement No. 2.

—. 2007. 5663rd Meeting (S/PV.5665) (Resumption 1), Provisional. New York, 17 April 2007.

—. 2011. 6479th Meeting (SC/10172). New York, 11 February 2011.

—. 2011. 6479th Meeting (S/PV.6479), Provisional. New York, 11 February 2011.

## Secondary Sources

Agrawala, S., A. Moehner, M. El Raey, D. Conway, M. van Aalst, M. Hagenstad, and J. Smith. 2004. Development and climate change in Egypt: Focus on coastal resources and the Nile. Environment Directorate: Working Party on Global and Structural Policies. Paris: Organisation for Economic Co-operation and Development.

Deligiannis, T. 2010. What are the essential elements of environmental security assessments? Poster session presented at ENVSEC-IES Workshop on Environmental Security Assessment Methodologies (November). Belgium.

FAO. 2003. *Trade Reforms and Food Security: Conceptualizing the Linkages*. Rome: FAO.

—. 2010. *Monthly News Report on Grains*, 65 (September).

Gleditsch, N. P., R. Nordas, and I. Salehyan. 2007. *Coping with Crisis: Climate change and Conflict; The Migration Link.* International Peace Academy (May).

Gorlick, B. 2007. Environmentally displaced people: A UNHCR perspective. Poster session presented at UN Headquarters. New York (May).

Homer-Dixon, T.F. 1994. Environmental scarcities and violent conflict: Evidence from cases. *International Security* 19: 5-40.

IASC. 2009. Climate change, food insecurity and hunger: Key messages for UNFCCC negotiators. Technical paper of the IASC Task Force on Climate Change, submitted by WFP, FAO, IFRC, OXFAM, WVI, CARE, CARITAS, WHO, and Save the Children (October).

—. 2008. Climate change, migration and displacement: Who will be affected? (October). Retrieved 17 April 2011 from http://www.unhcr.org/refworld/docid/4a2d189e0.html.

Intergovernmental Panel on Climate Change. 2007. Food, fibre, and forest products, and perspectives on climate change and sustainability. In *Climate Change 2007: Impacts, Adaptation and Vulnerability*. IPCC fourth assessment report. Geneva.

Millennium Ecosystem Assessment. 2006. *Ecosystems and Human Well-being: Current State and Trends*. Washington, DC: Island Press.

Stern, N. 2007. The impacts of climate change on growth and development. In *The Economics of Climate Change: The Stern Review*. London: HM Treasury.

World Bank. 2011. *World Development Report 2011: Conflict, Security, and Development.* Washington DC: The International Bank for Reconstruction and Development/The World Bank.

WBGU. 2011. World in transition: A social contract for sustainability; Summary for policy-makers. Berlin: German Advisory Council on Global Change.

Zoellick, B. R. 2011. Opening remarks on food prices/pre-G-20.15 February 2011. Retrieved 14 April 2011 from http://www.worldbank.org.tr

# CHAPTER TWO:

# ENVIRONMENT: IMPACTS OF GLOBAL CLIMATE CHANGE

# A New Ethical Approach to Environmental Protection: Ecocentrism

Kivilcim Ertan[1] and Birol Ertan[2]

## 1. Introduction

Although the history of environmental problems dates back to the environmental pollution experienced in Ancient Greek city states, environmental awareness in the contemporary sense has developed only in recent decades. People have developed awareness and ideas concerning the need for urgent measures to combat environmental problems, especially in cases such as the death of more than four thousand Londoners over a few days due to the high concentrations of pollution in the past century, or the leak of 40 tons of methyl isocyanine gas and other chemicals from a pesticide factory in Bhopal city, India, on the 3rd December 1984, resulting in the death of 18,000 people and the poisoning of more than 150,000. The development of environmental awareness has gained significant momentum since the environmental disasters of the 1970s.

We live in an era of global environmental problems that cannot be confined to the borders of a particular country, and where all actions, activities and policies for environmental protection must be reconsidered in the context of new ethical approaches. It is an irrefutable fact that ethical approaches play a determining role in the prevention of any kind of environmental problems—especially global environmental problems—and policies in all fields, primarily the energy sector, that lead to global environmental problems. This study aims to explore the role of ecocentrism, a new ethical approach that must guide decision-makers and citizens in environmental protection, and in preventing environmental problems and their effects in the process of energy production and consumption.

[1] Associate Professor, Lecturer, TODAİE, Ankara, TURKEY.
[2] Associate Professor, Lecturer, TODAİE, Ankara, TURKEY.

## 2. The Concept of Environment

Environment is the relationship of living and non-living things with each other and their surroundings. In this sense, the environment should be considered as the surroundings of life.

It should be kept in mind that humans are an inseparable part of the environment. The United Nations World Charter for Nature emphasises that mankind is a part of nature. It should be accepted that an environmental approach that focuses only on the environment in terms of the benefits for and happiness of mankind is both inconsistent and wrong in its approach towards the non-human environment.

### A. Environment

Although in daily life, the term "environment" is used synonymously with the concepts of "surrounding", "circle" or "entourage," in many disciplines the term covers a very large area, whose boundaries are unpredictable. In daily life, "environment" refers to everything except mankind. The field of environmental ethics cannot be limited to such a definition, and this opens up the possibility of considering the subject within the scope of environmental ethics. Defining the concept of "environment" within the context of environmental science and environmental ethics is of vital importance.

One common problem faced in defining environment is that the term may be taken to refer only to greenery and trees, or be limited to germination and tree planting, while environmental problems are considered within the context of the impacts of (for instance) littering. Such narrow approaches, which overlook environmental integrity, lack a comprehensive and consistent understanding of the environment based on ecological balance. According to Keleş and Hamamcı, who attempt to formulate a comprehensive definition of environment, in a general sense the term refers to the sum of physical, chemical, biological and social factors that might have a direct/indirect or immediate/non-immediate impact on human activities and living within a specific time period (1988: 25). Though this definition is considered an extensive definition, it can be accepted as satisfactory in terms of the emphasis it lays on the integrity of environment and the adoption of the ecological balance approach.

Turkish Environmental Law defines environment as the biological, physical, social, economic, and cultural surroundings in which living things maintain their inter-relationships and interact with each other throughout their life-cycles. This can be said to be a comprehensive definition that reflects the environment of life understanding.

Having reached a position on the scope and content of the concept of environment, we can begin to explore the subject of what environmental problems actually mean.

## B. Environmental Problems

According to Turkish Environmental Law, environmental pollution refers to all kinds of adverse developments in the environment that have a negative effect on the health of living beings, environmental assets, and ecological balance. Despite its comprehensiveness, it is hard to accept that this definition defines environmental problems adequately. Environmental problems comprise those biological, chemical, and physical impacts and activities that might adversely affect the health of mankind and other living things in air, water and land, or that have destructive effects on the ecological balance of a particular area or on the world as a whole. This definition can be said to be comprehensive and, to a large extent, inclusive.

Environmental problems, as understood under such a definition, are divided into several main types. The types of environmental problem include air pollution, water pollution, destruction of the earth, reduction in the diversity of plants and animals, noise pollution, and global environmental problems. Environmental problems of any type might cause others. Environmental problems do not respect national boundaries, and it is an undeniable fact that any one environmental problem may be the source of another. For example, over time, water pollution necessarily leads to soil pollution, especially as the result of agricultural activity.

## C. The Effects and Consequences of Environmental Problems

What are the adverse effects of environmental problems that threaten the health of human things, other living things, and ecological balance? First of all, environmental problems are cumulative. Some activities which do not affect human health over a short period may sometimes accumulate over time, and may come to have an adverse affect on the health of other people and living things. For example, immediately after the earthquake and tsunami in Japan, there was no immediate risk of leaks in radioactive material that might affect environmental health; however, there is now a very dangerous environmental problem, due to the increasing amount of radioactive leaks.

It is known that any kind of environmental problem can have a negative effect on the health, life and life quality of humans and other living things. It is also well-known that air pollution has deadly consequences, especially for people with respiratory tract problems and conditions such as asthma. In addition, long-term air pollution may lead to widespread respiratory and lung diseases, cause diseases such as nose and throat cancer, and lead to adverse and destructive effects on natural and architectural structures. Water pollution not only has adverse effects on living things, but also leads to soil pollution, resulting in other forms of pollution and associated damage. Many diseases that were very widespread in the past, such as typhoid, dysentery, jaundice and cholera are caused by water pollution. It is known that nitrates transmitted to the human body through polluted water can easily be converted into carcinogenic substances and thus cause cancer. Water pollution not only has negative impacts on people, but also results in death and harm to other living things—chiefly fish. Soil that is polluted due to agricultural pesticides, chemicals, bad fertilisation practice, overfertilisation, and hazardous waste may create immediate or gradual adverse effects on human health and the lives of other living things and plants. Soil pollution resulting from desertification and erosion leads to serious economic losses in the field of agriculture. Noise also has severe effects on the health of humans and other living things—it may cause psychological problems, hearing disorders, and difficulty in concentrating, thus resulting in workforce loss.

It is believed that the effects of global environmental problems will have a very destructive impact upon all living beings on earth, as well as on the ecosystem. Many countries may be submerged as a result of the melting of the polar glaciers due to the greenhouse effect, many health problems (chiefly skin cancer) may occur because of the increased levels of ultraviolet radiation caused by the degradation of the ozone layer, and there will be widespread loss of lives and property due to the floods and landslides caused by deforestation. Desertification will place the world's food reserves in jeopardy. The result of all this will probably be serious diseases, health problems, famine, poverty, and chaos. For all these reasons, the gradual increase and accumulation of environmental problems is considered to be a worldwide global crisis. In brief, we have to change our environmental viewpoint and ethical approach and to question the understanding that sees the environment as a freely consumable resource. It is clear that this is closely related to the issue of environmental awareness.

## 3. Environmental Awareness

Environmental problems are not a new phenomenon; they did not first appear in the last quarter of the 20th century, or after the Industrial Revolution. In fact, the history of environmental problems is as ancient as the history of human and environment relations itself.

It is known that Ancient Greek city states experienced environmental problems such as air pollution. By 312 BC, the Tiber River was so polluted that the Romans had to construct aqueducts to obtain clean drinking water. In 1236, lead pipes were used to transport water to London (Ponting, 2008: 417). As this example makes clear, environmental problems existed in the centuries before Christ, and a range of actions and approaches towards these issues were already being developed. Severe environmental problems, confronted in different areas of the world at different times in history, have brought about the development of environmental awareness at both a governmental and an individual level. It may be helpful to briefly examine this subject from a historical perspective.

### A. A Brief History of Environmental Awareness

Serious awareness of the relation between environment and human activity emerged no earlier than the second half of the 20th century. Faced with severe environmental problems, like the death of more than four thousand people due to air pollution in London in 1952, mankind has started to become more sensitive to the issue of environmental protection. In the years following the panic caused by the loss of more than 4000 lives due to air pollution in 1952, environmental awareness has developed at a national level, and an increasing number of people believe in the need to promote concrete measures at an international level. The United Nations Conference on the Human Environment can be considered the first example of the development of international environmental awareness on the issue of environmental protection.

As mentioned above, one important example of environmental awareness relates to the environmental disaster that occurred in Bhopal, India. On 3 December 1984, more than 40 tons of methyl isocyanine gas leaked from a pesticide plant in Bhopal, killing 18,000 people and poisoning more than 150,000. This terrible environmental disaster became instrumental in improving environmental awareness worldwide.

Today, environmental sensitivity has continued to develop in the face of the destructive effects of environmental problems, which are now so widespread that new legal regulations have been formulated across the world to protect the environment. Environment protection-oriented provisions have been included in constitutions and environmental protection organisations have been established. The development of environmental awareness in Turkey will now be examined.

## B. Significant Developments in Environmental Awareness

Following the fatal consequences and other severe effects of environmental problems, new ideas and approaches to the protection of environmental assets have begun to develop. The first serious example of such a new approach is the Meadows Report, commissioned by the Club of Rome.

In 1970, the Club of Rome commissioned Professor Donella Meadows of the Massachusetts Institute of Technology (MIT) to prepare a report on the prevention of the approaching environmental crisis. This report, which would come to be known as "the Meadows Report," was published under the name "Limits to Growth." The report argued that a major environmental disaster was awaiting mankind, unless ongoing rapid growth was halted.

The Stockholm Conference was another important development in environmental awareness at the international level. In 1972, an international environmental conference was held in Stockholm, Sweden. The conference was attended by representatives from 113 countries, 19 inter-governmental agencies, and more than 400 inter-governmental and non-governmental organisations. This Stockholm Conference was characterised by fierce debate. It was stated in Article 4 of the Declaration, published after the Conference, that most of the environmental problems in developing countries were caused by underdevelopment. The view that was accepted at the Conference, and expressed in its Declaration, was a formula proposed against the Zero Growth thesis, a net approach reflecting the "developmentalist approach" that prevailed until the last quarter of the 20$^{th}$ century.

Both the Zero Growth thesis and the idea that environmental problems are caused by underdevelopment (and that development is therefore inevitable) are examples of the anthropocentric approach to environmental problems. Not only are these approaches incapable of addressing environmental protection, but they are not adequate for the purpose of preventing environmental pollution.

Beginning in the 1980s, new approaches to environment and economy relations began to emerge in the context of increasing environmental awareness. Their purpose was to prevent environmental problems by regulating relations between humans and the environment. The introduction of the concept of Sustainable Development—in the report entitled "Our Common Future" and commissioned to the team of Norwegian former Prime Minister Gro Harlem Brundtland by The United Nations World Commission on Environment and Development in 1986—was an important development. This report was published in 1987. The understanding that environmental resources like water, soil, and air (considered as the source and limit of the economy and of economic development) should be preserved has led to the emergence of the understanding of "sustainable development". In the "Our Common Future" Report, a turning point in environmental awareness, sustainable development was proposed as the way of meeting the needs of the present without compromising the ability of future generations to meet their own needs (WCED, 1987). The concept of Sustainable Development thus represented a break from both the Club of Rome's Zero Growth Thesis and the understanding of development against all odds reflected in the Stockholm Declaration. Sustainable Development is an important step towards establishing a balance between environmental protection and economic growth.

In the late 20$^{th}$ and early 21$^{st}$ centuries, new approaches and thoughts on the protection of the environment have continued to be voiced, and new principles and suggestions have been proposed. The most important of these are the Preventative Approach and the Precautionary Principle. The preventative approach, based on the prevention of environmental problems before they arise, suggests that it is easier to prevent environmental pollution before it arises and that the cost of treating existing environmental pollution or reversing environmental conditions back to pre-pollution conditions is higher. The precautionary principle is based on the understanding that scientific uncertainty should be no excuse for failing to take action to prevent environmental problems. That is to say, the precautionary principle refers to taking necessary actions against activities that present an uncertain potential for significant harm, despite scientific uncertainty. The precautionary principle has not only been adopted as policy by EU countries, but in some countries (such as Germany) it constitutes one of the basic principles of environmental law.

## C. Environmental Awareness in Turkey

The evolution of environmental awareness in Turkey can be traced back to the legal arrangements made for the protection of environmental assets. Environmental regulations that manifested themselves in various legal documents were in time transformed into efforts to merge domestic (national) legal arrangements into a legal text in many countries, particularly Western countries. Nevertheless, a lot of countries have begun to enter provisions within their constititions enshrining the duties and responsibilities of the state and individuals for protection and development. The provision of Article 56 in the Turkish Constitution, related to "the right to environment", can be cited as an example of such a legal arrangement. Article 56 says: "Everyone has the right to live in a healthy, balanced environment. It is the duty of the state and citizens to improve the natural environment and to prevent environmental pollution." This provision imposes a legal duty or responsibility on the state and citizens for the protection of the environment. The Environmental Law No. 2872 of 1983 was amended by Law No. 5491, dated 2006. The new law aims to preserve the environment, the common asset of all living things, in the context of the principles of sustainable environment and sustainable development (Environmental Law No. 5491 of 26 April 2006, Article 1). While the Environmental Law—which has been revised a number of times since its first enactment in 1983—first adopted the principle of the indispensability of economic development, the concepts of sustainable environment and the understanding of sustainable development have, through recent amendments, become its imperative principles.

In Turkey, environmental awareness has not been limited to legal arrangements. The enforcement of environmental law for environmental protection purposes, other environmental legal regulations, the approval of international treaties towards the preservation of environmental assets, the enactment of numerous rules and regulations for environmental protection, and the establishment of the Ministry of Environment as an environment management agency, responsible for environmental protection (today named as Ministry of Environment and Urbanism), are other examples of environmental awareness that reflect upon implementation. Moreover, environmental protection-oriented jurisdictions, increased environmental awareness of citizens, the establishment of environmentalist non-governmental organisations, the increase in political parties' and non-governmental organisations' sensitivity to the environment, and the improvement of environmental education are other examples of improved environmental awareness.

The rise in environmental awareness in Turkey has occurred in line with its global development. Despite the gradual improvement in environmental awareness during the 1970s, it can be said that serious progress only started to be made during the 1980s. It has been asserted that environmental awareness saw its golden age after the 1990s. However, this golden age does not indicate that environmental awareness in Turkey has reached a satisfactory level. The reason for this is that despite progress at the legislative and discourse level, in practice there there remains a long way to go.

It is our approach to environment that guides our thoughts and actions in the process of both the development of environmental awareness and the implementation of policies for environmental protection. Our approach to the environment is directly related to environmental ethical approaches. Not only our thoughts, but also our actions, are guided and shaped by our ethical approaches. Therefore, the following section will examine the environmental ethical approaches that shape environmental protection thoughts and policies.

## 4. Environmental Ethical Approaches

The difference between the concepts of ethics and morality should be emphasised. While morality refers to our ideas and behaviour about what is good or bad, the purpose of ethics is to establish and defend the principles and rules that guide our behaviour. It is the philosopher's mission to justify these principles and rules and to explain why people must behave in accordance with them (des Jardins, 2006: 269).

Environmental philosophy is a broad field, covering epistemology, metaphysics, philosophy of science, history of philosophy, and normative fields such as ethics, aesthetics, and political philosophy (Jamieson, 2008: ix). Environmental ethics is one of the major fields of study in environmental philosophy. In this framework, studies conducted on ethics also contribute to studies in environmental philosophy. This section will focus on environmental philosophy in general, and on different approaches to environmental ethics in particular.

### A. Environmental Ethical Approaches and Protection of Environment

The concepts of Morality and Ethics are derived from the Latin and Ancient Greek words *mores* and *ethos,* meaning custom. However, although they are sometimes mistaken for each other, these two concepts have different meanings. According to Louis P. Pojman and Paul Pojman (1998: 4), morality refers to the customs, principles, and practices of a person or culture, and ethics refers to the whole domain of morality and moral philosophy. Accordingly, in this study, the term "ethics" (not "morality") will be used to refer to the principles that shape our behaviour, values, and thoughts towards the environment and that thereby determine our approach to it.

Three different ethical approaches to environmental ethics have emerged, where environmental ethics can be defined as the ethical principles involved in building environmental/human relations. These are anthropocentrism, biocentrism and ecocentrism.

### i. Anthropocentric Ethical Approaches

In this section, which examines environmental ethical approaches, anthropocentric ethical approaches that aim to preserve the environment only for the benefit and happiness of human things will first be discussed. One point should be underlined: even though anthropocentric ethical approaches (which may be classified as either weak or strong) are criticised by some philosophers, these notions must be analysed with care, since they are the first step towards the idea of environmental protection. Though anthropocentric ethical approaches are regarded as utilitarian and speciesist, or as a form of human chauvinism, they should be considered as an important phase in environmental awareness, in terms of the development of increasing interest in environmental issues.

One of the first anthropocentric ethical approaches to environmental protection was voiced by the Ancient Greek philosopher Epicurus. According to Türkaya Ataöv (2009: 68-73), Epicurus, starting from Jeremy Bentham's principle of "utility" based on utilitarianism, considered the limitation of desires, the practice of virtues, and the pursuit of happiness through friendship, as the process of understanding nature. Ataöv sees Epicurus as one of the first environmentalists. Epicurus' search for balance in the individual's relationship with nature can also be considered as one of the first examples of an anthropocentric ethical approach.

According to John Passmore, one of the defenders of anthropocentric ethical approaches to environmental protection, there is no need for a new environmental ethics; if the Western tradition of ethics is fully implemented, it can easily encompass the environment. Passmore voices two reasons for this, which conflict with each other. The first is that a being is useful as long as it serves human interest, and the second is that the importance attributed to humans relates to the extent to which they produce contemporary and cultural values (Passmore, 1974: 116). Passmore's anthropocentric approach is one of the best examples of the view that the environment must be protected in the interest of mankind.

It is very clear that today, anthropocentric approaches are inadequate for the purpose of conducting and regulating the environment and human relations. It should be kept in mind that the approach, which attempts to conduct and regulate the natural order that has emerged through an evolutionary process over billions of years through values and interests of the history of humanity dating back no more than a few thousand years, has today brought the world to the brink of environmental crisis. As a result, new approaches that go beyond the anthropocentric model have emerged.

### ii. Biocentric Ethical Approach

In the early 1970s, debates on the need for a new environmental ethics began, and the concept of inherent worth was frequently invoked. These debates resulted in the development of the biocentric approach, an example of a non-anthropocentric environmental ethics. "The biocentric outlook upon nature revolves around four central beliefs. First, humans are seen as members of earth's community of life, in the same sense and on the same terms as all other living things. Second, all species, including humans, are part of a system of interdependence. Third, all living things pursue their own good in their own way (the "teleological centre of life" belief). Finally, humans are understood as not being inherently superior to other living things" (des Jardins, 2006: 139).

Biocentricism, which has laid the foundation for the ecocentric perspective that sees nature as assets created not only for humans, and that widens the domain of interest to other nonhuman species, has created important opportunities for presenting and criticising the inadequacies of the anthropocentric viewpoint.

One of the pioneers of biocentric ethics is the Nobel Prize winner, environmentalist, and animal lover, Dr. Albert Schweitzer. According to Schweitzer, all living things have inherent worth and deserve our love and respect. Many books and articles have been written about Schweitzer, who is considered to be among those who laid down the foundations of biocentricism. Another is Paul Taylor, who has made significant contributions to the biocentric viewpoint. In his book *Respect for Nature,* Taylor discusses biocentric ethics, and defends a "form of life" view, according to which the interactions between the members of the ecosystem community are strengthened based on an ethical bond. According to Taylor, humans, as the sole ethical practitioners in the Earth's biosphere, are ethically responsible for limiting their adverse environmental impacts, in order to eliminate the destruction of their living surroundings (Taylor, 1986). As des Jardins writes, "Paul Taylor's 1986 book *Respect for Nature* offered one the most fully developed and philosophically sophisticated contemporary defences of a biocentric ethics. Taylor's approach explains the normative claim that all living beings have a good of their own and we have duties towards them as required by the concept of inherent worth." (des Jardins, 2006: 136)

One idea which constitutes the foundation of biocentric ethics concerns animal rights. According to Peter Singer,—one of the first philosophers to put forward a conception of animal rights based on Jeremy Bentham's theory of utilitarianism—although many philosophers have advocated equal consideration of interests in one form or other as a basic moral principle, few have recognised that the principle has applications beyond our own species. One of the few who did so was Jeremy Bentham (Singer, 1988: 49).

In the literature, debates concerning human rights are often conducted via a comparison between the views of Peter Singer and Tom Regan. Regan (1988: 40) states that three goals shape his thought on animal rights: the total abolition of the use of animals in science, the total dissolution of the commercial animal agriculture system, and the total elimination of commercial and sport hunting and trapping. In his 1983 book entitled "The Case for Animal Rights," Regan criticised utilitarianism based on a Kantian philosophy of ethics that takes the idea of rights as a starting point. He argued that animals, like humans, have inherent rights (Jamieson, 2008: 116-117). On Regan's approach (1983: 243), rights do not cover all rights (as with human rights), but include the inherent right to live, the right to freedom, and the right not to be exposed to torture and violence.

Advocates of the biocentric view are not limited to those who defend animal rights. There are also philosophers who support the right to existence of plants, as living things. For example, Goodpaster (1978: 308-325) argued that plants also have rights.

Christopher Stone maintains that plants (especially trees) have rights, like other living species. Stone published a book in the mid-1970s in which he defended the right of trees to exist (this was based on his law review article, arguing for the preservation of the trees in the Sierra Nevada of California, where Walt Disney Enterprises had been permitted to build a leisure complex). In 2010, Stone's book was revised and republished. Stone (2010: xi) complained in the introduction that while the book had been published 35 years previously, there had as yet been no judicial development concerning the rights of trees.

To sum up, the biocentric outlook, which focuses on the right to live of non-human living things (such as animal and plant communities) constitutes a viewpoint based on the life of living beings, in terms of environmental ethics. According to this approach, humans are not superior to nature but are themselves part of the community of living things. It is worth noting that this outlook does not attribute any worth to non-living things.

### iii. Ecocentric Ethical Approach

Like other approaches, the literature on animal rights in particular (and on biocentric ethics in general) has its problems and weaknesses. One primary problem is the radical approach to animals, which we use as food. Such seeming contradictions pose potential difficulties for the biocentric outlook, in relation to our behavioural and legal norms. Hence, some writers have begun to develop ecocentric approaches.

The ecocentric ethical standpoint argues that not only humans and other living organisms, but also non-living things (especially the ecosystem), should be included in the sphere of interest of ethics. In his book on land ethics, Aldo Leopold (1968: 224), one of the leading environmentalist philosophers of the 20th century, defended the ecocentric ethical approach, writing that "a thing is right when it tends to preserve the integrity, stability, and beauty of the biotic community."

One of the most important foundations of ecocentrism is the ecological viewpoint and the conception of the integrity of nature. An important contribution to this approach is the Gaia hypothesis, developed by Lovelock and Margilus. The Gaia hypothesis maintains that earth is a living organism; if we cannot preserve Gaia, it is not possible to preserve the components of water, air, soil, and living beings. Many writers who have conducted studies on environmental ethics in recent years—primarily the English scientist James Lovelock and the American biologist Lynn Margilus—have proposed that life on earth can be considered a living organism (Lovelock, 1979; Sagan and Margilus, 1983).

In the present discussion, environmental ethical approaches have so far been classified under the headings of anthropocentric, biocentric, and ecocentric approaches. There are also new and somewhat different approaches that fall outside this classification, or that challenge this classification while still potentially being included within it. Among the philosophical views that cannot be classified as anthropocentric, biocentric or ecocentric are the following: deep ecology, sociological ecology, ecofeminism, and vegetarianism.

How can the ethical approaches described so far guide us in preserving the environment, and what role can they play in environmental protection?

The next section will focus on the effects of environmental ethical approaches, especially in relation to energy.

## 5. The Ecocentric Ethics in Energy Production and Consumption

The basic concern of contemporary environmental ethics is to regulate the deteriorating relations between human beings and the environment. All environmental ethical approaches (with the exception of the individualistic and developmentalist perspective of classical economy, that treats the environment as a free and consumable resource), especially ecocentric ethics, endeavour to guide human behaviour and actions in order to prevent the destructive impacts of the global environmental crisis we face in the 21st century.

It is an indisputable fact that the majority of human activities that pollute the environment and upset the ecological balance have their origin in industrial activities, especially the process of energy production and consumption. Solutions to local and global environmental problems depend upon abandoning classical and against all odds-development approaches, and instead adopting new ethical approaches that prioritise environmental awareness. Among these approaches, the ecocentric outlook is realistic, feasible and oriented towards preventing global environmental crisis: it takes the integrity of the environment into account.

The viewpoint that sees environment and natural assets only as a source of economic development reflects the understanding of "economic development against all odds" recognised at the 1972 Stockholm Conference. Yet we have to comprehend that natural resources like water, soil and air (the source of economy) have their limits too. If we deplete or destroy these living resources, we will not be able to find another world in which to live.

In light of these considerations, the concept of sustainable development or growth has been proposed. The concept or idea of "sustainable development" is a rational environmentalist alternative, which has been propounded against the understanding of development against all odds, especially against a conception of the environment as merely a free natural resource to be consumed.

According to the entropic standpoint, energy cannot be transformed infinitely; some is always lost during energy transformation. Eventually, it runs out. When the law of entropy is applied to social and economic problems, very radical results are achieved. No issue reflects the law of entropy more clearly than the "energy crisis." Industrialisation and affluence have become possible with the operation of coal, oil and natural gas reserves; these resources have supplied fuel to electric power plants, factories, automobiles, airplanes, etc. These fuels are fossil fuels, derived from the decomposition and compression of prehistoric organisms, and are not sustainable; they cannot be replaced once they are used up (Heywood, 2010: 282). Whether or not we support the entropic viewpoint, it is clear that the world is gradually marching towards a severe energy crisis. Energy consumption is increasing, yet we continue to ruin the environment and natural resources at a rapid pace, and to pollute the environment in order to boost energy consumption.

The development of ideas and efforts towards establishing a sustainable balance between environment and economic development continues in every field, primarily in the field of energy production and consumption. One of these developments is the approval of an international environmental treaty. The Kyoto Protocol, an international step towards addressing global warming and climate change, aims to limit or reduce the emission of carbon dioxide that leads to climate change, as well as limiting five other greenhouse gases. The Kyoto Protocol treaty was negotiated in Kyoto, Japan, at the end of 1997; it was opened for signature in 1998, took final form on 15 March 1999, and entered into force in 2005. The countries that ratified the Kyoto Protocol have agreed to reduce their emissions of the six main greenhouse gases to 1990 levels. The Kyoto Protocol, today endorsed by 160 countries to cover more than half of greenhouse gases, aims to cut GHG to the level of 5%. The protocol is an important international legal document that supports the search for alternative energy resources, and sets forth realistic suggestions and measures for implementation.

The sustainability of energy policies is of crucial importance, if we are to meet the energy requirements of future generations. However, the current perception of the sustainability of energy policies suggests a sustainable energy supply. An energy policy that attaches priority to non-renewable resources like fossil fuels—or to long-term renewable resources—is far from being sustainable. Moreover, although sustainable alternative energy resources are known, investment in such resources has been limited, especially in developing countries. It is hard to speak of the sustainability and safety of energy policies in these countries, including Turkey (Uğurlu, 2006: 18). These are examples of the ongoing prevalence of anthropocentric ethics in the formation and implementation of energy policy. If we do not adopt ecocentric behaviour and actions in every field, including energy production and consumption, the measures being taken against the current global environmental crisis will remain inadequate.

## 6. Conclusion and Evaluation

It is a fact that our ethical perspectives shape our behaviour over the course of time. But our ethical perspectives manifest themselves not only in our views and behaviour, but also in the functioning of the institutions that shape social order and, above all, in positive law regulations (i.e., in the norms that give direction to social order). While constitutions, laws, and other legal arrangements (ethics is also reflected in local jurisdictions) are among the sources of positive law at the national level, this approach should find a place in the treaties and documents formulated at the international level.

The functioning of the ecosystem, which constitutes a chain process consisting of living and non-living beings that complement each other, does not consider humans and other beings as independent processes. Therefore, neither the anthropocentric outlook (that favours the understanding of nature's exploitation in the interest of mankind) nor the biocentric outlook (that reflects the living things-oriented understanding) can be favoured in terms of developing an environmental ethics capable of preventing environmental crisis.

Environmental protection should be considered as a whole; we should not overlook the fact that environmental pollution knows no boundaries, and that global environmental problems pose a threat to the whole world and to the life of all living things. What, then, is the resolution? The only possible resolution is to approach the environment as a whole, from individual behaviour patterns to institutional relations, from state administration to international governance processes, and from national or international rules that regulate the state and social life, and to incorporate ecocentric ethics into all these fields.

In conclusion, in the 21$^{st}$ century, when the earth will undergo a global ecological crisis and face the deadly effects of environmental problems that will gain increasing momentum in every country and every day, the environmental ethical viewpoint comes to the forefront as the most consistent, feasible, and realistic approach. It prioritises the integrity of the environment, and this promises hope for the future. If we do not change our thoughts and behaviour, our individual and institutional activities, and our management systems, and if we do not adopt an integrated approach for the sake of the protection of environment in general and the ecological balance of the earth in particular, more severe environmental crises are inevitable in the years to come. Today, the intercontinental effects of a nuclear plant located in Japan not only confirm the integrity of earth, but also prove that the world constitutes an ecosystem and that environmental problems know no boundaries. The solutions and policies needed to combat the deadly global environmental problems awaiting us can be developed by taking timely measures, from an ecocentric viewpoint.

## References


Armstrong, Susan J. and Botzler Richard G. (2004). Environmental Ethics: Divergence and Convergence. New York: McGraw Hill.

Ataöv, Türkaya (2009). Kapitalizm ve Çevre. İstanbul: İleri Yayınları.

Des Jardins, Joseph R. (2006). Çevre Etiği: Çevre Felsefesine Giriş (Çev. Ruşen Keleş). Ankara: İmge Kitabevi Yayınları.

Goodpaster, Kenneth K. (1978). On Being Morally Considerable. Journal of Philosophy, Sayı 75, 308-325.

Heywood, Andrew (2010). Siyasi İdeolojiler (Çev. Ahmet Kemal Bayram, Özgür Tüfekçi, Hüsamettin İnaç, Şeyma Akın, Buğra Kalkan). Ankara: Adres Yayınları.

Jamieson, Dale (2008). Ethics and the Environment. Cambridge: Cambridge University Press.

Keleş, Ruşen ve Hamamcı Can (1988). Çevrebilim. Ankara: İmge Kitabevi Yayınları.

Keleş, Ruşen, Hamamcı, Can and Çoban, Aykut (2009). Çevre Politikası. Ankara: İmge Kitabevi Yayınları.

Leopold, Aldo (1968). A Sand County Almanac. New York: Oxford University Press.

Lovelock, James (1979). Gaia, A New Look at Life on Earth. Oxford: Oxford University Press.

Lovelock, James (1988). The Ages of Gaia: a Biography of Our Living Planet. Oxford: Oxford University Press.

Passmore, John (1974). Man's Responsibility for Nature: Ecological Problems and Western Traditions. London: Duckworth.

Pojman, Louis P. ve Paul Pojman (ed.). (1988). Environmental Ethics: Readings in Theory and Application. Belmont: Wadsworth Cengage Learning.

Ponting, Clive (2008). Dünyanın Yeşil Tarihi, Çevre ve Büyük Uygarlıkların Çöküşü (Çev. Ayşe Başçı). İstanbul: Sabancı Üniversitesi.

Regan, Tom (1988). The Radical Egalitarian Case for Animals, Louis P. Pojman, Paul Pojman (ed.) in Environmental Ethics: Readings in Theory and Application (ss. 83-90). Belmont: Wadsworth Cengage Learning.

Rifkin, Jeremy (2010). Avrupa Rüyası (Çev. Buket Okucu Özbay). İstanbul: İstanbul Bilgi Üniversitesi Yayınları.

Sagan, Dorian ve Margulis, Lynn (1983). The Gaia Perspective of Ecology. Ecologist, 13 (5), 160-167.

Singer, Peter (1979). Equality for Animals? Practical Ethics. Cambridge: Cambridge University Press.

Stone, Christopher D. (2010). Should Trees Have Standing? Oxford: Oxford University Press.

Taylor, Paul W. (1986). Respect for Nature: A Theory of Environmental Ethics. Princeton, N.J.: Princeton University Press.

Uğurlu, Örgen (2006). Türkiye'de Çevresel Güvenlik Bağlamında Sürdürülebilir Enerji Politikaları. Ankara Üniversitesi Sosyal Bilimler Enstitüsü Yayımlanmamış Doktora Tezi.

WCED, World Commission on Environment and Development (1987). Our Common Future. Oxford, New York: Oxford University Press.

Wilkinson, David (2002). Environment and Law. New York: Routledge.

# Voluntary Carbon Trade: A New Livelihood Model for Low Income Families

İlker Met[1] and Taner Altunok[2]

## 1. Introduction

The increase in greenhouse gas emissions due to industrialisation has accelerated global climate change, which is moving rapidly towards a situation where its consequences will be irreversible.

As a matter of fact, it is estimated that since 1751, carbon dioxide ($CO_2$) emissions into the atmosphere have amounted to 337 billion tons (Boden, Marland, & Andres, 2010). The most important factor in this increase is the anthropogenic effect of heavy resource use (Onozaki, 2009). As a result, temperature levels have increased, meteorological abnormalities are encountered more frequently and severely, and the number of habitable places on earth is continuously decreasing. Even though there are studies stating that $CO_2$ level should be kept below 350 PPM (Hansen, et al., 2008) in order to sustain the earth as a liveable place, atmospheric $CO_2$ levels being at about 391 PPM today shows that we are in real contradiction. In short, our world needs to cool down fast. There are many alternative proposals, such as storing $CO_2$ in geological formations and using artificial trees to reduce carbon emissions; however, given that factories cannot be closed down in the present era, and technology that will quickly destroy $CO_2$ at the point of emission from a stack does not exist, the only way of preventing $CO_2$ emissions is to generalise those activities that reduce emissions. To bring this about, a global carbon strategy should be developed worldwide, and the necessary steps to bring

[1] PhD, Project Manager, Corporate Management Directorate, TR Ziraat Bank, Tandoğan Ankara, TURKEY, imet@ziraatbank.com.tr

[2] Professor of Engineering, Director, Graduate School of Natural and Applied Sciences, Çankaya University, Öğretmenler Caddesi No: 14 06530 Balgat Ankara, TURKEY, taltunok@cankaya.edu.tr

about the shift from a carbon community to a green community must be taken (Figure 2-1).

Within this general framework, many studies have been conducted in order to ensure the cooperation of all countries. For instance, the United Nations Environment Programme (UNEP) increased awareness by publishing a Climate Competitiveness Index in April 2010 (CCI, 2010). But these kinds of studies are in their early stages, and need to be generalised.

Figure 2-1 Annual Carbon Emission and Carbon Public Relationship

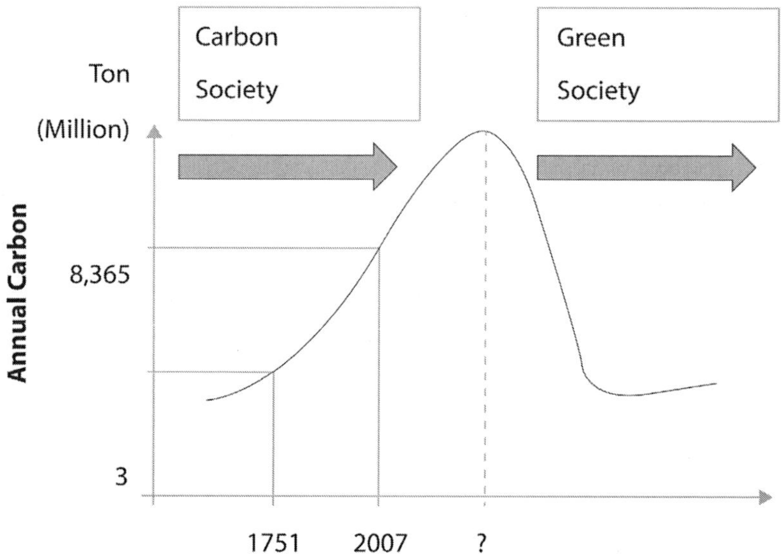


## 2. Basic Tools and Definitions

Since industrialisation and the shift from conventional energy resources to fossil fuel, anthropogenic effects have led to the gradual increase in pressure on the greenhouse gas emission absorption and assimilation capacity of the atmosphere. Accordingly, a shift to a green community has become essential. Achieving this shift depends on the possibility of replacing fossil fuels with alternative sources. For example, the British Government aims to reduce carbon dioxide emissions by 80% before 2050. The climate target of the EU, which states that "temperature should be kept below 2° C compared to pre-industrial levels" aims to reduce carbon dioxide emissions (Fouquet, 2010).

Despite many financial instruments and regulatory mechanisms being pursued in order to accomplish the targets for reducing carbon emissions, there is as yet no consensus. One possible precaution is a kind of taxation called $CO_2$ or energy taxing, which aims to control carbon emissions produced through human activities. Taxation on environmental pollution has two advantages. The first is the improvement of the environment; the second is economic efficiency, as the environmental tax revenues reduce other taxes that have an adverse effect on the labour supply and money saving decisions. On this basis, environmental taxes are being accepted as desireable, since they yield extra benefits for the community (Gimenez & Rodriguez, 2010).

From the perspective of global climate change, it is more desirable to accomplish production from beginning to end where $CO_2$ emissions are lowest, and only subsequently to conduct international commerce. In this way, global carbon emissions could be reduced. But given actual conditions, factors should not be taken into account on a purely economic basis. If production costs increase for carbon reduction measures, producers will lose their competitive advantage on international markets. Thus, it is important to develop appropriate implementations like Border Taxation Arrangements, quotas, or a consumption based greenhouse gas emissions inventory (Dong, Ishikawa, Liu, & Wang, 2010). Although there is no general consensus regarding such implementations at the international level, some progress is being made. Some European countries (Denmark, Germany, Finland, Norway, Sweden, Holland and Britain) have achieved green taxation reforms for 20 years at a national level. Since 2005, the EU has applied a scope Emission Trading Scheme (ETS). This is a good start, but it cannot be applied to the sector which is responsible for 50% of greenhouse gas emissions. Besides, energy taxes are very low in non-EU countries. This represents a significant difficulty, and limits the effect of tax mechanisms (Gimenez & Rodriguez, 2010).

Another instrument, the Carbon Stock Market, emerged from the Kyoto Protocol. This is a development of a fiscal instrument that was formerly used in combating air pollution in the US. The Carbon Stock Market started as a means of buying and selling carbon; thus, the carbon trade, also known as the Carbon Market, has become a method of controlling greenhouse gas emissions. The main motivation is the economic/monetary value given to carbon as a result of its contribution to air pollution. Thus, a company or country which reduces its greenhouse gas emissions to a level that falls below the target value can sell its extra allowance to another country or company, and can actually make a profit. The carbon trade was worth 30 billion dollars in 2006. The fact that it had reached 143.7 billion dollars in 2009 (Kossoy & Ambrossi 2010), despite the financial crisis that began in 2008, shows that it is likely to continue to develop steadily in the future.

Emission operations have been taking place in climate stock markets in a variety of countries, such as Australia, the US, and within the European Union. In this kind of stock market, futures on emission and climate derivatives with option contracts are processed. Despite debates concerning stock exchange market trading, issues like climate and the environment affect everyone, so it is thought that developing the stock exchange market as a tool for the development of the carbon trade is founded on a solid basis. So long as damage caused to the environment is not turned into financial value, the environment will continue to be polluted. The delicate balance here is that a mechanism should be established which works towards the elimination of environmental damage, rather than solely for the purpose of profit-making.

Another instrument is the Voluntary Carbon Market (VCM), which is the subject of the present study. VCM, which can be defined as the summation of all carbon credit transactions in non-compulsory markets, is based on the reduction of greenhouse gas emissions through the selling of non-compulsory credits or voluntary balancing credit to voluntary last users, not to compulsory buyers (CBCMM, 2011). Companies, firms and establishments buy carbon credits which are produced by Voluntary/Verified Emission Reductions (VER) Projects, in order to reduce or balance emissions through their activities.

The first voluntary carbon investment occurred when AES paid an amount equal to 50 million dollars to Guatemalan farmers for tree planting, in order to stabilise $CO_2$ emission. This payment was not made due to legal obligations, but within the marketing and promotion strategy of the company.

Green Goal was developed for the 2006 FIFA World Cup, and is still practiced today. It is one of the most popular examples of this process from the recent past. Within the scope of Green Goal, various measures have been taken to reduce the carbon emissions that can emerge as a result of the planning, construction and implementation of the facilities built to host a World Cup. Grant support was provided for renewable energy and energy efficiency projects in developing countries (Hochfeld & Stahl, 2003). The "Keidanren Voluntary Action Plan"[3] was applied in Japan, and

[3] http://www.keidan ren.or.jp/

a "Greenhouse Friendly Initiative" was applied by the Australian government.[4] These are all examples of systematically conducted programmes.

As can be seen, the most important element for the purpose of setting up this mechanism based on volunteer work is the willingness of participating companies to improve their image and to create an environmental brand identity. If this consciousness and awareness of the issues involved in Voluntary Carbon Markets spreads to the community, then the supply-demand balance is successfully impacted. In fact, as sensitivity to global climate change increases within the framework of social responsibility, companies, institutions, organisations and individuals are increasingly likely to use this market model.

Another advantage of the voluntary carbon trade is that there is no limit to participation, and emission reductions approved by many different standards can be requested by different buyers.

These advantages are also reflected in the numbers. According to Ecosystem Marketplace and Bloomberg data, the total trading volume in carbon markets has reached 93.7 Mt $CO_2$. Despite a 26% decrease in trading volume as compared to 126.6 Mt $CO_2$ in 2008, it still maintains a level 39% higher than the level in 2007. Due to the 2007 global financial crisis, this decrease mainly originates from cuts to projects that depend upon corporate social responsibility initiatives (especially carbon equalisation). However, the most important reason is the uncertainties regarding the status of the draft laws in the United States, which were expected to increase demand in these markets (ÇOB, 2011). Despite all of this, there is still hope that the market will gradually expand over the coming years (Oppenheim, 2009). In parallel to these projections, it is possible to envisage that voluntary markets will become a very important mechanism in the near future.

## 3. Current Situation Analysis for Turkey

At the 7th Conference of the Parties within the United Nations Climate Change Framework Agreement, held in Marrakesh, Morocco in 2001, Turkey's name was deleted from Annex II, and the participating countries were invited to accept special conditions regarding Turkey that distinguish it from those countries in Annex I. Turkey was the 189$^{th}$ country to sign the agreement on May 24, 2004 (ÇOB, 2010). Turkey participated in the

[4] http://www.climatechange.gov.au/greenhousefriendly/

Conference of the Parties, in subsidiary meetings and in negotiations on equal terms with others, since it is still a party to UNCCFA.

Since the Kyoto Protocol brings heavier regulations and burdens determining binding emission reduction obligations, based on the UNCCFA Annex-1 list, Turkey was finally able to become party to the Kyoto Protocol in 2009. Because Turkey was not a party to the Protocol in 2005, when the Protocol was put into force, it was not included in Annex B, where the greenhouse gas reduction responsibilities of the participating countries are listed (Atamer, 2010).

Since Turkey participated in Annex I in the framework of UNCCFA, it cannot host Clean Development Mechanism (CDM) projects, a flexibility mechanism created within the scope of the Kyoto Protocol. Similarly, since Turkey is not on the Annex B list within the scope of the Kyoto Protocol (and takes no responsibility for reducing greenhouse gas emissions numerically), it cannot organise Joint Implementation (JI) projects either. Thus, the Voluntary Carbon Market is the most flexible mechanism that can be used by a country like Turkey (i.e., one that has no obligation under the Kyoto Protocol to reduce greenhouse gas emissions) (Öztürk, 2009). Turkey has participated in international efforts on climate change only recently, so necessary proficiency from its institutions on the topic of the environment is just beginning (Atamer, 2010). The number of VCM projects registered so far in Turkey amounts to 108. 50% of these projects relate to hydroelectric plants, 42% to wind energy, 5% to waste storage, and 3% to geothermal projects. According to the 2009 VER data, the annual market volume is 81.6 million, while the equivalent of an 8 million ton $CO_2$ reduction has been achieved. Turkey's share of the Volunteer Carbon Market is around 6%. On the markets, Turkey's Emission Reduction Certificates are exchanged at half or one quarter of the rate of certificates of other countries that trade within the scope of the Kyoto Protocol (Khan, 2010). These figures show that Turkey produces below its potential. As a matter of fact, the 64 projects countrywide have involved a loss of 300 million US dollars. As regards emission reduction projects, this is largely due to the lack of an evaluation mechanism, a database, and an institutional organisation for the regulation of the emission certificate market (Atamer, 2010). However, within the scope of the "Increase of Capacities in Turkey to Cope with Climate Change Project," preparatinos for legal regulation continue. On the other hand, the existence of the aims of energy and carbon emission markets (established in the "Istanbul International Financial Centre Strategy and Action Plan," published in 2009) is an indication that Turkey has started to develop infrastructure studies concerning this issue. If we consider that energy-related carbon emissions are expected to reach 210 million tons by 2020 (Yüksel & Sandalcı, 2011), it is clear that Turkey is likely to benefit from voluntary carbon trading, if the necessary systems are created.

## 4. What to Do?

Even though Turkey is a country with potential, its to-do list is very long. It is possible that Turkey could become powerful in terms of its carbon trade as a result of systematically formed policies. It is estimated that the most appropriate policy for Turkey to follow is the organisation of "voluntary carbon markets as a tool for social responsibility, increasing business efficiency and competitiveness." In order to develop the proposed strategy, policies should be strengthened in the four main areas listed below.

1. Waste recycling
2. Improvements in energy productivity
3. Increase in usage of renewable energy resources
4. Supporting projects that aim to reduce carbon emission

In the market economy, regulatory policies should be formed and applied by governments with respect to carbon trading. The main policy objectives should be to obtain a legal basis for the carbon market, and to increase transaction volume. In order to achieve these targets, greenhouse gas emissions and emission reduction should be recorded, and the traded carbon certificates should have a standard. So, two keywords ("carbon footprint" and "energy efficiency") should take their place at the heart of the development of carbon policies.

In order to make Volunteer Carbon Markets work, it is very important to construct VER projects. As climate change awareness increases, the demand for these projects will increase. Thus, in order to balance carbon emissions, the people or institutions responsible for these emissions should provide financial support for renewable energy, land use and forestry, methane capture, and waste recycling projects that lead to the reduction of greenhouse gas emissions, as this is necessary for the functioning of the system (Öztürk, 2009). However, the creation of projects is not enough. With the help of these projects, the precise reduction in carbon emissions should be determined through a neutral institution: in other words, a certification mechanism should be established. So, within the framework of the Turkish Standards Institution (TSE), a centre of carbon emissions should be established, and measurement standards should be formulated.

The first worldwide standard for VCM was published in 2006, and standards are developing gradually. In the development of standards for voluntary carbon markets around the world, VCS (Verified Carbon Standard), Gold Standard, and VER + have been used more often than others (Öztürk, 2009). Procedures need to be defined concerning which standard to use in Turkey, and how this is to comply with standards used around the world. Emissions created by institution should be measured by a responsible ministry, such as the Ministry of Environment and Forestry. Thus, by creating carbon standards, these will spread across the country.

Energy efficiency should not be the direct target of the implemented policies. As a matter of fact, as Van den Bergh (2011) mentions, energy saving has no effect as a replacement for environmental regulations—as in the case of $CO_2$ emissions pricing the effect on environment. Energy saving is only effective if environmental regulations are already in place. Thus, energy saving policies should emerge in parallel with the necessary legal environmental regulations.

Another important consideration is that all firms should adopt the "Carbon Management Rules" to be developed by the Ministry of Environment and Forestry. Taking the relevant legal regulations into account, the institutions quoted on the Istanbul Stock Exchange (IMKB) should harmonise institutional management principles with carbon management principles. As a result, awareness will be created in firms, and this should canalise support for projects that reduce carbon emission.

The creation of a physical market where carbon can be bought and sold is another necessary tool of the system. Execution of this process by a public organisation which designates strategy will make it easy to put a national climate strategy into practice. The presence of public organisations will create a natural guarantee mechanism for the technical reliability of financed projects. The Turkish Derivatives Exchange (TURKDEX), which started to operate on the 4th of February 2005 in İzmir, could be used to construct this market. This kind of application has already been proposed, in Arı's (2010) study.

In order to develop the Volunteer Carbon Trade, another important point concerns the development of financing models. Considering that various international organisations are primary buyers of carbon certificates in VCM (as in the case of the 2006 World Cup), these kinds of supports are important resources that can create attraction centres. Incentives offered by the government and the support of banks (which have an important role in the development of markets) are very important for projects that aim to minimise carbon emissions. It is important to position public organisations as voluntary buyers during the development of the carbon market. In this context, voluntary carbon trading as a business model for low-income families presents itself as an opportunity.

## 5. Carbon as a Livelihood Model for Low Income Families

Even though, at first sight, reduction in carbon levels can be classed as a high cost project and deemed an unrealistic dream, many such projects are successfully applied in the world today: these include the integration of low income families into the banking system through the M-PESA project (Hughes & Lonie, 2007), low income families getting access to computers through the production of 100 dollar laptops through the OLPC project (OLPC, 2005), and bank credits for low income families developed by Muhammad Yunus (Nobel, 2006). This shows that the usage of carbon trade is a possible opportunity for low income families too.

In fact, even though the carbon trade is the new livelihood model for low income families in developing countries, and many projects have been created for its use in the implementation of development targets around the world over the last two years, standards and legal arrangements are not yet fully formed. Considering the fact that many low income families do not own houses or land, the question of these families can benefit from the carbon trade will become a new field of study in the near future.

In the present study, a proposal namely "tree fields" is explained, in order to mark a starting point for Turkey. As Table 2-1 shows, forestation activities have increased rapidly since 2006. The General Directorate of Forestry Strategic Plan and the National Climate Change Strategy Paper, which set important targets for forests, shows that there are serious development prospects for increasing the role of forests in carbon capturing. In fact, Khan (2010) considers the 181.4 Mt $CO_2$ carbon capture potential (which corresponds to the reforestation and rehabilitation of 3 million hectares) shows the very positive potential for the development of a carbon market in the forestry sector. What has to be done is to establish a mechanism to link this potential to the carbon market.

**Table 2-1 Forestation Studies by Years (AGM, 2011)**

| | 2003 | 2004 | 2005 | 2006 | 2007 | 2008 | 2009 | 2010 |
|---|---|---|---|---|---|---|---|---|
| **Total** | 116,929 | 165,343 | 177,705 | 401,168 | 400,196 | 463,592 | 501,387 | 483,493 |

As can be seen from the activity steps in Table 2-1, the first step for the creation of the tree fields is to determine suitable locations and suitable tree species. Since low-income families will operate the tree farms, these farms should be located in relatively poor regions (from a development level perspective), and contain a mixture of land suitable for plantation. In determining tree species, the added value of the by-products of the trees should be considered. In order to make progress on this issue, the Ministry of Environment and Forestry and Scientific and Technological Research Council of Turkey (TUBITAK) should fund university research projects. After determining the location and tree species, TSE should measure the amount of carbon that will be absorbed by tree fields and issue a certificate. This will help to calculate tree field areas and the value of the certificates needed by institutions.

There are different approaches to determining the value of certificates. The first is to determine the values of certificates by the Turkish Derivatives Exchange, which buys and sells commodities. This approach has the advantage of allowing VCM to develop rapidly. However, speculative movements valid for each derivative present a significant threat to the carbon market. Another approach is the determination of the values of carbon certificates by responsible public institutions, by making the certificates subject to processes of buying and selling. Making the buying of certificates obligatory for those institutions (or even individuals) that emit carbon will make trade in these certificates easier. Whichever method is applied, the literature shows clearly that significant value is to to be gained from tree fields. For example, (McHalea, McPhersonb, & Burkea's (2007) research shows that the starting value of certificates can be determined, assuming that a tree cleans ~0.8 ton $CO_2$ annually. Considering that in a 50-year period, a tree provides oxygen at about 31,000 USD, air pollution control at about 62,000 USD, water treatment at about 37,500 USD, soil erosion control at about 31,250 USD and prevents alternative costs (CTC, 2011), it can be seen that secondary factors are also important in the determination of certificate values. The most important points to address are the need to create a pricing mechanism that will determine the added value from the operation of tree fields, and the need to convey this added value to low income families (not using carbon certificates for speculative reasons).

The amount of financial support needed during the establishment of tree fields may be calculated by Ziraat Bank, which has experience in the field of agricultural credits. Furthermore, financial support for the credits proper to investment can be given by the Bank. Once these steps are taken, the mechanisms for selling of carbon absorption certificates of tree fields in the carbon market will be established. By the time these certificates are available to trade on the derivatives exchange, companies aiming to reduce their carbon emission will buy them and find the opportunity to balance their carbon emissions. Evaluating these certificates based on fixed terms will provide companies with the opportunity to balance their increasing carbon emissions during intensive production periods.

**Table 2-2 Necessary Action Steps for Tree Fields and Responsible Institutions**

| Action Steps | Responsible Institution |
|---|---|
| **1.** Determining appropriate lands and tree species for tree fields | TR Ministry of Environment and Forestry, TUBITAK, Universities |
| **2.** Certification of the carbon absorption capacity by the tree fields | Turkish Standards Institution (TSE) |
| **3.a.** Creation of market and derivatives to provide the trade of certificates | Turkish Derivatives Exchange (TURKDEX) |
| **3.b.** Determination of the certificate values by responsible institutions | Ministry of Environment and Forestry |
| **4.** Selling the operational rights of tree fields for 10 years period | Ministry of Environment and Forestry |
| **5.** Providing necessary financial support for tree fields | Ministry of Environment and Forestry<br>Ziraat Bank |
| **6.** Periodical certificate standard control of the tree fields | General Directorate of Forestation and Erosion Control |

The most important action needed for the trade of tree field certificates is to guarantee the amount of carbon held by these certificates. The most significant problem is the maintenance cost of the fields during the early years. Many trees die within the first year of plantation due to disease, dehydration, and human effects. Another problem is the crowded population living in and around the forests. However, these negative effects will be removed and will become positives through giving operational responsibility to low income families, because citizens living in rural areas will gain income from the trade in certificates as a result of the operation of tree fields. In this way, tree fields will be protected, and an important income transfer mechanism for development will be established. While the transaction cost of working with a low income population will increase because of inadequate carbon market infrastructure, start-up capital will be financed by funds from experienced banks for projects approved by the Ministry of Environment and Forestry.

## 6. Results

As prevailing energy production methods change, finding more efficient energy resources must become a major new economical orientation. In order for this to occur, there is a need for a niche market which is ready to pay more. Global climate change can create niche markets, through the process of transforming new ideas for the prevention of carbon emissions into economic value. VCM is at the centre of this cycle. Consumers choose the most economical options, so the development of this niche market can be made possible through state incentives. Although, on the basis of past experience, transformation to a low carbon economy is expected to be quite slow (Fouquet 2010), it is predicted that the carbon trade will develop much more quickly in the future, given increasing global awareness of climate change.

In this context, sustainable development management might become more effective through spreading global climate management principles throughout society, and developing the voluntary carbon market in Turkey. Funding socially responsible institutional projects, especially in areas like renewable energy, energy efficiency, land use and forestry, methane absorption, limiting the storage of organic materials, and using waste as raw material, will all have a key role to play in achieving these goals.

Tree fields (amongst other projects) have the potential to become an important model for Turkey. Turkey falls behind Europe in terms of its forested areas. However, through carbon certificates, tree fields, and the development of carbon sinks, increasing the forest area by forestation and the improvement of the carbon market will become possible. Besides, through giving operational responsibility for tree fields to low income families, a national partnership for development will be established. By implementing the actions summarised in this study, a tool for Turkey's participation in the global carbon market after 2012 can be developed. Hence, low income families, who are the most vulnerable to climate change, will be protected. Furthermore, by taking the carbon market away from the monopoly of big projects, the carbon trade will spread across a larger market. As a result, a step on the way towards the removal of the negative impacts of global climate change will have been taken.

# References

AGM. (2011, 04 08). Ağaçlandırma Genel Müdürlüğü. 04 08, 2011 (access) Statistics: http://www.agm.gov.tr/AGM/Files/istatistikler2009/2ormanTesisiCalismalari.xls

Arı, İ. (2010). İklim Değişikliği ile Mücadelede Emisyon Ticareti ve Türkiye Uygulaması. Ankara: DPT.

Atamer, S. A. (2010). İklim Değişikliği Politikaları Mevcut Durum Değerlendirmesi Raporu. Ankara: Çevre ve Orman Bakanlığı.

Boden, T., Marland, G., & Andres, R. (2010). Global, Regional, and National Fossil-Fuel $CO_2$. Oak Ridge: Carbon Dioxide Information Analysis Center.

CBCMM. (2011, 04 08). TC Çevre ve Orman Bakanlığı. 04 08, 2011 (access), Glossary: http://www.karbonkayit.cob.gov.tr/Karbon/Files/terimlers%C3%B6zl%C3%BCg%C3%BC.pdf

CCI. (2010, 04 01). The Climate Competitiveness Index. 04 13, 2011 (access) Retrieved from http://www.climatecompetitiveness.org/index.php?option=com_content&view=article&id=19&Itemid=10

CTC. (2011, 04 10). Colorado Tree Coalition-Benefits of Trees in Urban Areas. 04 11, 2011 (access) http://www.coloradotrees.org/benefits.htm

ÇOB. (2011). Karbon Piyasalarında Ulusal Deneyim ve Geleceğe Bakış. Ankara: TC Çevre ve Orman Bakanlığı.

ÇOB. (2010). Ulusal İklim Değişikliği Strateji Belgesi. Ankara: TC Çevre ve Orman Bakanlığı.

Dong, Y., Ishikawa, M., Liu, X., & Wang, C. (2010). An analysis of the driving forces of $CO_2$ embodied in Japan–China trade. Energy Policy, 38, 9784-6792.

Fouquet, R. (2010). The slow search for solutions: Lessons from historical energy transitions by sector and service. Energy Policy, 38, 6586-6596.

Gimenez, E. L., & Rodriguez, M. (2010). Revaluating the first and the second dividends of environmental tax reforms. Energy Policy, 38, 6654-6661.

Hansen, J., Sato, M., Kharecha, P., Beerling, D., Berner, R., Masson-Delmotte, V., et al. (2008). Target Atmospheric $CO_2$: Where Should Humanity Aim? Open Atmospheric Science Journal, 2, 217-231.

Hochfeld, C., & Stahl, H. (2003). Green Goal–Environmental objectives for the 2006 FIFA World Cup. Berlin: Öko-Institut.

Hughes, N., & Lonie, S. (2007). M-PESA: Mobile money for the "Unbanked" turning cell phones into 24-hour tellers in Kenya. Innovations, 63-81.

Khan, M. A. (2010). Türkiye'nin Ormancılık Sektörü ve Karbon Piyasası Raporu. Ankara: TC Çevre ve Orman Bakanlığı.

Khan, M. A. (2010, 03). United Nations Development Programme. 04 03, 2011 (access) Carbon Markets Strategy. Setting the stage for Turkey: http://www.undp.org.tr/energEnvirDocs/Carbon_Markets_Strategy-Setting_the_Stage_for_Turkey_(Amin_Aslam_23032010).pdf

Kossoy, A., & Ambrossi, P. (2010). State and the trends of the carbon market 2010. Washington: Carbon Finance at World Bank.

McHalea, M. R., McPhersonb, E. G., & Burkea, I. C. (2007). The potential of urban tree plantings to be cost effective in carbon credit markets. Urban Forestry & Urban Greening, 6, 49-60.

Nobel. (2006, 10 13). Nobel Price. 02 11, 2011 (access) http://nobelprize.org/nobel_prizes/peace/laureates/2006/press.html

OLPC. (2005). One Laptop per Child. 02 11, 2011 (access) http://laptop.org/en/vision/index.shtml

Onozaki, K. (2009). Population Is a Critical Factor for Global Carbon Dioxide Increase. Journal of Health Science, 55 (1), 125-127.

Oppenheim, J. (2009). The Future of Carbon Markets. Londra: World Coal Institute.

Öztürk, M. (2009, 06 02). Mustafa Öztürk. 03 30, 2011 (access) Gönüllü Karbon Ticareti: http://www.mozturk.net/Upload//karbon.pdf

Tans, P. (2011, 02). NOAA Earth System Research Laboratory. 04 02, 2011 (access) Trends in Atmospheric Carbon Dioxide: http://www.esrl.noaa.gov/gmd/ccgg/trends/

Van den Bergh, J. C. (2011). Energy Conservation More Effective With Rebound Policy. Environ Resource Econ, 48, 43-58.

Yüksel, I., & Sandalcı, M. (2011). Climate Change, Energy, and the Environment in Turkey. Energy Sources, 33, 410-422.

# ANALYSIS OF TURKISH ENVIRONMENTAL ENERGY POLICIES IN A REGIONAL FRAMEWORK*

ÇIĞDEM ÜSTÜN[1]

In the 1970s, environmental depredation and the exhaustion of natural resources were identified as two important threats, leading to the redefinition of national security (Dikmen, 2004: 31-32) to include environmental security as a top priority topic in the agendas of states, international organisations, and national and international civil society organisations. Simultaneously, energy security, particularly the security of energy supplies, became a national security issue, due to the oil crisis of 1973. The oil embargo imposed by oil producing Arab countries resulted in high oil prices, which forced energy consumer countries to reconsider their energy policies and relations with OPEC members. As OPEC member states increased the price of oil, the life blood of European industry and the diversification of energy routes and resources became vital. Therefore, since the 1970s, it can be observed that easy access to energy resources, diversification of routes, and smooth transit passage of oil and natural gas from energy producing countries to energy consumer countries has become an important subject in security debates within the international system. In this context, the significance of transit countries and seas has increased, since these are the key geographical areas in this desired secure energy transition. However, as the oil and natural gas transit

* An earlier version of this paper was presented as *"Türkiye'nin Çevre ve Enerji Politikalarının Uluslararası Düzlemde Analizi"* at I. International Symposium on Global Climate Change, Environment and Energy-Global Challenges & Opportunities to Global Stability, Istanbul Aydin University, 25 April 2011, Istanbul, Turkey and it was published in Çiğdem Üstün, "Turkey Between Environmental Protection and Energy Security: A Regional Perspective", *Insight Turkey*, Vol. 14, no 2, 2012, pp.177-192.

1 Associate Professor, Chair, Political Science and Public Administration Dept., Vice Dean, Faculty of Economics and Administrative Sciences, Gediz University, TURKEY, cigdem.ustun@gediz.edu.tr

from the Middle Eastern, Mediterranean and the Black Sea countries to western European countries has amplified, environmental concerns and threats have increased as well. In other words, states, international organisations, companies and individuals have started to prioritise two controversial security issues—environmental and energy—at the same time. This debate has spread all over the world, but has most affected the transit areas in the Mediterranean and the Black Sea.

Research has shown that international attention given to the Mediterranean Sea has generated more positive impacts on the protection of its environment, as compared to that of the Black Sea (EU4SEAS, 2010). Industrialisation around the Black Sea during the Cold War, the lack of international attention over many decades, and the region's post-Cold War position as a crucial hub for the transport of the energy produced by Caucasian and Black Sea littoral countries to the energy consuming countries in Europe, have all aggravated the environmental situation in the region. Pollutants created by chemical industries and oil leaks from tankers have led to a decrease in biological diversity. Thus, the increasing pollution of the seas has entered the agendas of governmental and non-governmental international/regional organisations, and individual states in the last two decades. Unfortunately, after the end of the Cold War and the collapse of the Soviet Union, the main priorities of the newly independent states included neither an increase of biological diversity nor a decrease in pollution. As the regional states put their efforts into competing on the international liberal markets, they focused on increasing industrialisation and on establishing trade and economic ties with the energy demanding countries. With these aims, both energy demand and supply security became the top priorities of the region, forcing environmental issues into the background, although international and regional agreements were discussed and signed.

Since the 1970s, the United Nations has been working hard to increase awareness of environmental protection and to emphasise the need for international cooperation. There have been several agreements at the international level which have set rules for states to follow in order to protect the environment. However, these UN rules and regulations are fundamentally based on voluntary compliance, with no serious sanctions facing those states that choose not to comply. Therefore, states enter these agreements if they are beneficial to their own interests, or in order to show their good intentions, rather than because they believe in the importance of the purpose of the rules and regulations (Bilen et al., 2008). Since energy and environmental issues are perceived as political matters, the harmonisation of energy and environmental policies and the application of these accepted rules and regulations have become problematic. Worse still, as political relations and conditions of the region change, these agreements alter (EU4SEAS, 2010).

Since the Chernobyl disaster in the Black Sea region, it has been understood that environmental problems are not restricted to their countries of origin, and that environmental threats are as important as the conventional threats of the Cold War era. The Chernobyl power reactor accident of 1986 forced Gorbachev to include ecological security in his Perestroika policies in 1987. After the collapse of the Soviet Union, economic stabilisation, development, energy security and trade stood out in the debates in the region. In the same way, international society started to discuss environmental security in the 1990s, which forced the countries around the Black Sea to emphasise energy security, economic development, and environmental issues. Therefore, a vibrant relation between energy and environment policies became an indispensable relation for the Black Sea region as a whole. Particularly after the term "sustainable development" entered into discussions in the Black Sea region after the UN Summit in 1992 in Rio (Ülengin et al., 2010: 844), it has been observed that the regional countries have started to debate the need for economic development without degrading the environment. However, this relation has proved to be problematic, since energy and the environment have been perceived as two conflicting policy areas. The desire of individual states to increase their economic benefits, especially by increasing energy based trade with energy consuming countries, has estranged them from the sustainable development understanding.

Industrial activity in Russia, Ukraine and Belarus causes air pollution, due to the proliferation of chemical and nuclear industrial power plants. Also, in these countries, pollution in the soil prevents productive agricultural activities. On the basis of his ecological research in the Black Sea, Laurence Mee defined the region as an environmental catastrophe (Mee, 2002). Eutrophication/Nutrient enrichment, chemical pollution (including oil), changes in marine living resources, overfishing, biodiversity changes, and climate change negatively affect the lifestyles and living standards of the people living in the region. Unfortunately, individual states continue with their industrialisation projects while disregarding the significance of sustainable development.

At the international level since the 1990s, smaller regional and bigger international organisations have started to implement projects in order to emphasise the importance of environmental threats and to counteract the environmental damage perpetrated by the industrial and energy sectors. There have been several European Union (PHARE, Black Sea SCENE,

Black Sea Global Ocean Observing System (Black Sea GOOS), SESAME, PLANCOAST, TACIS/EUROAID and so on) and UN (GEF/UNDP Black Sea Environmental Programme) funded projects and programmes, which differ from each other in terms of their priorities (EU4SEAS, 2009). But they all share the aims of increasing awareness and of helping to create new control mechanisms. They have all made a small but positive impact in increasing biological diversity and decreasing environmental pollution.

As cooperation has increased since the end of the Cold War, the Black Sea shows signs of recovery, at least to a certain extent. However, more needs to be done regarding the assessment of basic environmental threats, the state of the environment and the price of inactivity, the development of an environmental policy, the adoption of legal documents and ratification, and the implementation and enforcement of those legal documents. There is an urgent need to deal with energy security, economic development and environmental matters concurrently, both at the national/regional and international levels.

A rapid increase is being observed in the need for energy resources. Developed countries need 75 per cent of the energy supply (Omer, 2008: 2268) due to increased industrialisation, changes in lifestyles, and increased levels of migration from rural to urban areas. It is projected that until 2030, yearly energy necessity will increase 1.8 per cent (Abdeen, 2008: 2272) and that natural gas necessity will increase 2.3 per cent. Also, the need for natural gas for industrial use will increase by up to 47 per cent by the same year (Bilen et al., 2008: 1532). The International Energy Agency predicts that the increase in energy need will reach the level of 60 per cent, and it is foreseen that this kind of increase in energy demand may not be met by the energy producing countries alone (Luciani and Mazzanti, 2006). With renewable energy use at only 14 per cent (Bilen et al., 2008: 1534), there is an urgent need for better utilisation of alternate types of energy resources. Studies demonstrate that decentralisation of the production of biomass, sun, wind and geothermal energy (Bilen et al., 2008) and the construction of infrastructure for the distribution of these types of environmentally friendly energy resources are necessary. Unfortunately, concerns over economic development, rivalry among regional states, and the lack of an international and national desire to construct renewable energy producing power plants and infrastructure for distribution of this type of energy prevent international and regional organisations taking steps in this direction.

Although it is known and agreed at the international level and in regional organisations that cooperation between economically developed and less developed countries is necessary (i.e., the UN Millennium

Declaration), political conflicts (hot and cold) hamper cooperation at both regional and international levels. The lack of cooperation and agreement among the regional states increases the number of tankers passing through the Black Sea, and rivalry regarding oil and natural gas pipelines decreases the effectiveness of the pipeline projects, resulting in the continuation of environmental threats for the sea in general, and for the Bosphorus and Dardanelles in particular.

Research has shown that biological diversity has decreased considerably, not only in the Black Sea but also in the Bosphorus and Dardanelles (Uysal et al., 2002). Even the migration routes of the birds in the region have changed, and the diversity of sea birds passing through the region has decreased (Birpınar et al., 2002). Unfortunately, all these negative effects of the increased traffic of tankers in the region have not raised awareness of environmental issues to necessary levels. Although international and regional organisations try to focus on these threats, political conflicts prevent regional states acting together and complicate possible cooperation efforts (Çelikpala, 2010).

At the international level, governmental and nongovernmental organisations pay particular attention to the control mechanisms for tanker traffic and the renovation of tankers and harbours (i.e., International Marine Organisation, Helsinki and Bucharest Agreements, UN Sea Law, and so on). However, these efforts have not fully prevented leaking tankers passing through the sea, since agreements at the international level have not triggered amendments in the national Sea Law of the regional states, or compelled the states to implement the international law vigorously.[2] Again, rivalry in industrialisation and economic development among the regional states is observed as the main reason for ecological problems (Carpenter et al., 2005).

The Black Sea is an isolated sea. Therefore, chemical pollution, oil leakages, and decreases in biological diversity are observed with greater frequency than in the Mediterranean and the Aegean seas, thus hampering tourism activities in the Black Sea basin (Üstün, 2009). Although tourism has been an important economic activity, increasing political, social and economic relations among the states and the people, as well as competition in the energy field combined with political conflicts, have once more decreased the significance given to environmental policies and the related tourism activities in the Black Sea riparian states.

---

[2] Bonn Agreement, 24 November 2010, Dublin, Ireland
http://www.bonnagreement.org/eng/

Today, research and studies demonstrate the need for the construction of alternative energy transit routes and new diverse energy resources in the wider Black Sea region (Richard and Heard, 2005). Unfortunately, instability, imbalance, and conflicts among states have a negative effect on levels of cooperation, trade and security in the region. The lack of trust among these states prevents almost all types of economic, social and political cooperation, including those related to the pipeline projects. In fact, the researchers and representatives of the regional states and organisations agree that balanced regional development is necessary, in line with international agreements, for the sustainable development for all the states around this particular sea basin (EU4SEAS, 2009).

In the field research conducted in the Black Sea countries (Turkey, Greece, Bulgaria, Moldova, Ukraine, Georgia, and Armenia) researchers, practitioners, and politicians emphasise that the environment cannot be separated from geopolitics and geopolitical issues, which are without stability, cooperation and economic development (EU4SEAS, 2010). Experts have identified Nagorno-Karabakh (between Armenia, Turkey and Azerbaijan), Abkhazia and South Ossetia (among the South Caucasus countries) as the main problem areas in the region, and have stated that it is wishful thinking to believe in environmental cooperation at the international and/or regional level unless political solutions are obtained for these conflict situations (EU4SEAS, 2010).

As one of the biggest countries in the region, Turkey has a potentially major role to play in environment and energy related policies and cooperation efforts. After the collapse of the Soviet Union, Turkey initiated some regional cooperation mechanisms and actively participated in others. However, like the newly independent countries, Turkey concentrated on economic development and energy policies, and became a part of regional conflicts: as a result, environmental concerns were shunted, even though Turkey is one of the main victims of environmental degradation in the region. The negative effects of pollution and the decrease in biological diversity in the Black Sea affected Turkey both economically and socially. More importantly, tanker traffic in the Bosporus and the Dardanelles, which amounts to 2000 tankers every year carrying around 250 million barrels of oil (EU4SEAS, 2010), creates a big risk for Istanbul and the other industrial cities in the region. In this framework, Turkey started to search for alternative ways to transfer oil and natural gas from the energy producing regional countries to (mainly) Europe, since the increase in the demand for energy in the EU countries opened a new and profitable opportunity for Turkey to boost its economic development. Turkey, situated right in the middle of energy producing and consuming countries, perceived this geopolitical position as a window of opportunity, and focused on reliable, diverse, and cost effective energy policies, while aiming to create a more liberal energy market. However, like other regional countries, Turkey has a demand-oriented rather than an efficiency-oriented energy policy. Although an energy efficiency law was passed in 2004, demand orientation has been the main priority (Turkish Ministry of Foreign Affairs, 2009). In this framework, Turkey wants to benefit from all the natural gas and oil resources of its neighbours, and therefore to implement a strategic energy policy in its region (Babalı, 2009). Population increase, rising economic needs, technological developments and increasing energy demands in relation to these factors occupy an important place in Turkey's economic, political, and energy policies (Ozturk et al., 2004). Since the 1990s, Turkey's energy production has increased by 4.3 per cent, but this increase has brought with it an increase in carbon dioxide emissions. Research shows that Turkey's carbon dioxide emissions are set to increase by 5.8 per cent every year until 2025 (Bilen et al., 2008: 1557).

The increase in energy production has not kept pace with the energy demands of Turkey's industry and people. Therefore, like the EU countries, Turkey is dependent on energy imports, mainly from its neighbours. In the 1980s, Turkey started to take steps for the construction of oil and natural gas pipelines, and since the 1990s it has been observed that the infrastructural investment for these pipeline projects has accelerated (Hacısalihoğlu, 2008: 1868). Projects such as Samsun-Ceyhan, Baku-Tbilisi-Erzurum, NABUCCO, and Burgaz-Alexandropolis aimed to increase the volume of the energy carried, but also to bypass the Black Sea because of environmental concerns. However, although the regional countries are dependent on each other as regards energy supply-and-demand security, environmental concerns and economic relations, political disagreements, rivalries and conflicts still continue to have a negative effect on these projects (i.e., Burgaz-Alexandropolis pipeline) (Tsakiris, 2011).

Regional countries which are not able to overcome these conflictual relations do not pay sufficient attention to renewable energy resources. The increasing significance of natural gas and the dependence of industrial production on natural gas has hindered the construction of renewable energy infrastructural projects since the early1990s (Bilen et al., 2008: 1546).

Although international agreements deal with environmental protection, increasing biological diversity and pollution (sea, air and soil), imbalances regarding economic development, underdevelopment of good governance, and political and military conflicts between the states and unconsolidated democracies in the region all create major obstacles in the implementation of these international agreements (EU4SEAS, 2010). However, cooperation among civil society organisations and municipalities between Greece and Turkey, Georgia and Armenia, and Romania and Turkey demonstrate that environmental cooperation and the implementation of the agreements are achievable goals. Research has shown that as people living in the polluted areas realise the negative effects, people-to-people contacts will trigger cooperation (EU4SEAS, 2010). However, at the governmental level, energy is perceived as a high politics area which prevents the better implementation and application of environmental agreements, rules, and regulations.

Since the environment is perceived as an area that is closely related to energy policies, high politics sees pipelines for oil and gas as a means of decreasing tanker traffic in the Black Sea, Bosphorus and Dardanelles, and thus as a more environmentally friendly alternative. However, pipelines are as dangerous as tankers for the people living along their routes. The protection of the pipelines is a vital concern, due to the political and military conflicts among different groups in the region. Pipelines are also very vulnerable in cases of natural disaster such as earthquakes, flood, and fire. Existing pipelines and those being constructed (Blue Stream, White Stream, NABUCCO, BTC and Samsun–Ceyhan) are significant for both the EU and Turkey in terms of energy supply security. However, interruptions in these pipelines create problems for electricity production in EU countries. Especially since the 1990s, natural gas has been used for electricity production (Say, 2006) and in recent years, interruptions of natural gas transfer, particularly from Russia to Ukraine, have demonstrated that electricity is not a sustainable energy type either.

Another alternative that has been debated is nuclear energy. Increasing population, developing economies, and increasing energy needs push countries towards finding cost effective energy resources. In this framework, nuclear energy has become a topical issue due to its investment possibilities. In order to save itself from dependency on the energy producing countries, Turkey has once again started to debate nuclear energy power plants (Kaygusuz, 2010: 1077). The possible construction of nuclear power plants in Sinop and Mersin has been discussed for a very long time in Turkey; however, licensing problems in Sinop left Mersin-Akkuyu as the only alternative location. It has been foreseen that these power plants, to be realised with Russia, will produce 5000 megawatt energy, and that the investment will cost around 20 billion US Dollars (Kılıç, 2008). This project, while creating an important energy resource for Turkey, will also increase economic and trade relations between Turkey and Russia (CNNTURK, 2011). Nevertheless, nuclear power plants also create threats for the people and the region—some similar to the threats presented by pipelines, others unique to nuclear energy. After the earthquake in Japan in March 2011, nuclear threat has been the number one topic in energy debates, forcing EU member states to revise their nuclear energy policies, and the opposition's hand against the nuclear power plants has become more powerful. The debates have focused on using renewable energy resources rather than nuclear energy, since accidents (i.e. Chernobyl) and natural disasters (i.e. Japan) endanger people's lives due to radioactivity. However, changing state agendas, relations, and priorities all affect public opinion. Therefore, it is not currently possible to predict the future of nuclear energy plants (Bilen et al., 2008).

Renewable energy resources are accepted as the best solution to the problems of environmental protection. Water, sun and wind are seen as the main energy resources in the fight against environmental threats, when used alongside modern technology. However, we need to be careful with these energy resources as well—especially water—since hydroelectric power plants have an effect on natural and social environments. For this reason, there is a need for vigilant calculation in the construction of these plants (Bilen et al., 2008: 1537). On the other hand, solar and wind power are especially favourable for countries in the Mediterranean region, including Turkey.

After the 1990s, the emphasis on renewable energy resources in Turkey increased, but the number of investors in this sector was limited. The new Renewable Energy Law is expected to increase both the number of investors and the volume of investment (Bilen et al., 2008: 1547). In Turkey, research has demonstrated that there is a need for investment not only in the construction of renewable energy power plants, but also in the distribution of wind and solar energy. As long as the distribution of the energy produced by wind tribunes and solar panels is not realised, the demand for these energy resources will be kept at minimum levels. The same applies to geothermal water resources in Turkey. Some villages and districts of big cities in the Aegean region use geothermal water, especially for heating purposes. However, the lack of investment and infrastructure for this type of energy limits its wide utilisation. In Turkey, although research conducted since 1962 demonstrates that the volume of geothermal energy is 4500 megawatts (Bilen et al., 2008: 1555), it is clear that new infrastructural investments are needed for this volume of energy to be used.

Hydroelectric power is an important energy resource for Turkey. SAP, with 22 damns and 19 hydroelectric power plants, is the biggest power plant in the region. Social, economic, and natural effects on the environment need to be studied carefully, and new investments should be encouraged in line with the results of these investments. It is believed that hydroelectric energy would be an important investment for the country, along with geothermal water resources and sun. Solar energy is often used in hotels and summer homes in Turkey, but power plants generating energy from solar panels are seldom seen. The Aegean and the Mediterranean coasts of the country, using solar panels for heating, could also use wind power to generate 83 gigawatt wind energy (Bilen at al., 2008). Unfortunately, significant investment trends are not being observed in this sector either.

Turkey's recent signing of the UN Framework Convention on Climate Change should be perceived as an important first step in improving its record on air, water and soil pollution, since air pollution controls in most Turkish cities have not been taken as seriously as in Ankara and some districts of İstanbul. Within the same framework, water pollution should be prevented by increasing cooperation between the littoral states and decreasing sea traffic. Even since the signing of the Kyoto Protocol in 2004, emission values have not improved. Although the new EPDK (Energy Market Regulatory Authority-EMRA) law allows investors to realise energy generating projects up to 10000 megawatts, the lack of infrastructure for the distribution of the energy produced limits the utilisation of environmentally friendly renewable energy sources.

## Conclusion

Although the environment has been one of the main debating issues in the international arena, energy demand and the security of energy supplies have overshadowed environmental protection. Especially after the late 1980s and the beginning of the 1990s, the need for energy in the energy consuming countries rose, due to increased industrialisation. As states and people put their efforts into reaching the easiest and cheapest energy resources, environmental concerns were shunted aside. However, recent natural disasters, the increased threat to human security, and concerns over the sustainability of energy sources have combined with environment debates to spark discussions on sustainable development, both in economic terms and with regard to energy policies. The same debates and concerns have also been observed in the Black Sea region as a whole, but the political and sometimes military conflicts among the regional states, combined with rivalries in energy and economic policies, prevent the regional states from cooperating effectively and focusing on sustainable development. Although renewable energy resources are being discussed at the regional and international governmental level and by non-governmental institutions, states continue to perceive energy policies as part of high politics. Therefore, natural gas, oil and nuclear energy resources are seen as foreign policy tools in the hands of the states in the region. It can be argued that before the political and military conflicts come to an end, and democratisation and good governance are established in the regional states, environmental concerns will continue to be discussed, although little appreciated.

## References

Babalı, T., "Turkey at the Energy Crossroads", *Middle East Quarterly*, 2009, pp. 25-33, http://www.meforum.org/2108/turkey-at-the-energy-crossroads

Bilen et al., "Energy production, consumption, and environmental pollution for sustainable development: A case study in Turkey", Renewable & Sustainable Energy Reviews, 12, (2008), pp. 1529-1561.

Birpınar et al., "Environmental effects of maritime traffic on the İstanbul Strait", *Environmental Monitoring and Assessment,* 152 (1-4), 2002, pp. 13-23.

Carpenter S. R. et al. (2005). Ecosystems and Human Well Being, Millennium Ecosystem Assessment, 2, Island Press, http://pdf.wri.org/ecosystems_human_wellbeing.pdf

CNNTURK, Akkuyu Nükleer Santrali için geri sayım, 16.03.2011, http://www.cnnturk.com/2011/dunya/03/15/akkuyu.nukleer.santrali.icin.geri.sayim/610087.0/

Çelikpala, M., *Security in the Black Sea Region Policy Report II*, Commission on the Black Sea, (2010). http://www.blackseacom.eu

Dikmen, Ç. "Enerji ve Çevre Ekseninde Ulusal Güvenlik", Görüş, (September 2004), pp. 31-32 http://www.maden.org.tr/resimler/ekler/43de7cf7ba769c7_ek.pdf

EU4SEAS Project, *International Seminar on Environmental and Maritime Issues in Europe's four sea basins*, Reykjavik, Iceland, May 2009.

EU4SEAS Project Database, Energy and Transport Work Package, Turkey, Greece, Bulgaria, Moldova, Ukraine, Azerbaijan, Armenia and Georgia Interviews, 2010.

Hacısalihoğlu B. "Turkey's natural gas policy", *Energy Policy*, 36, 2008, p. 1868.

Kaygusuz, K., (2010)_Sustainable energy, environmental and agricultural policies in Turkey, *Energy Conversion and Management*, 51, p. 1077.

Kılıç, S. (2008), İlk nükleer santral Mersin'e yapılacak, NTVMSNBC, 13.02.2008, http://arsiv.ntvmsnbc.com/news/435457.asp

Luciani, G. and M. R. Mazzanti, "Italian Energy Policy: The Quest for More Competition and Supply Security", *International Spectator*, 3, 2006, pp. 75-89.

Mee, L. D., *Protecting the Black Sea Environment: A Challenge for Co-operation and Sustainable Development in Europe*, Centre for European Policy Studies and International Centre for Black Sea Studies, 2002.

Omer, M. A., "Energy, environment and sustainable development", *Renewable & Sustainable Energy Reviews*, 12, 2008.

Ozturk et al., "Residential–commercial energy input estimation based on genetic algorithm (GA) approaches: an application of Turkey", Energy and Buildings 36, 2004, pp. 175-183.

Richard, J. P. and J. Heard, "European Environmental NGOs: Issues, Resources and Strategies in Marine Campaigns", Environmental Politics, 14: 1, 200, pp. 23-41.

Say, N. P., "Lignite-fired thermal power plants and SO2 pollution in Turkey", *Energy Policy*, 34, 2006, pp. 2690-2701.

Tsakiris, T., Burgas-Alexandroupolis-death of a great pipeline project? EKEM European Energy Policy Observatory, (2011). http://www.ekemeuroenergy.org/en/index.php?option=com_content&view=article&id=173:burgas-alexandroupolis-death-of-a-great-pipeline-project&catid=35:analyses&Itemid=57

Turkish Ministry of Foreign Affairs, Deputy Directorate General for Energy, Water and Environment (2009), Turkey's Energy Strategy, http://www.mfa.gov.tr/data/DISPOLITIKA/EnerjiPolitikasi/Turkey's%20Energy%20Strategy%20(Ocak%202009).pdf

Uysal et al., "Benthic Community Structure of the Bosphorus and Surrounding Area", *Water Science and Technology,* 26 (8), 2002, pp. 37-44.

Ülengin F. et al., "A problem-structuring model for analysing transportation-environment relationships". European Journal of Operational Research, 200(3), 2010, p. 844.

Üstün, Ç., *An overview of environmental and maritime issues in the Black Sea,* International Seminar on Environmental and Maritime Issues in Europe's four sea basins, EU4SEAS Project, Reykjavík, Iceland, May 2009 http://www.eu4seas.eu/images/stories/projects/publications/reykjavik/eu4seas_cigdem_ustun.pdf

# EUROPEAN UNION ENVIRONMENTAL DIRECTIVES AND A PRACTICE ON BUSINESS

MERAL EROL FIDAN[1]

## 1. Introduction

In Turkey, preparation towards harmonisation with EU directives and legislation is continuing. Nevertheless, not much is done about enterprises' environment and waste management. About 300 directives of the *acquis communitaire* directly relate to environmental issues, and 14 of these directly regulate waste management. Only two of these directives have a binding effect on member states, but candidate states must adapt their legislation in accordance with the other directives.

Industrial Zones (IZs) are Turkey's way of promoting industrialisation within the planned development concept. These zones are established in order to found industrial areas in suitable places, in order to prevent uncontrolled industrial development and environmental problems, and to use resources rationally. Industrial zones, which lower infrastructure costs and revive the local and regional economy, are effective tools for industrial development. From this point of view, industrial zones offer numerous economic and social benefits for society.

In terms of environment, Industrial Zones (which gather industrial enterprises as a result of correct choices) support planned urbanisation and regional development through infrastructure and service opportunities. On the other hand, industrial zones may also have dangerous effects on the environment and create health and security threats, as a result of choosing the wrong location, the wrong applications, and through the lack of adequate supervision.

[1] Assistant Professor, Bilecik Şeyh Edebali University, TURKEY

## 2. European Union Environmental Directives

All legislation concerning the environment that must be followed by the European Union member states are in “Directive” format. There may be variations among member states with regard to the application of these directives, due to differences in domestic laws and forms of government. Not all member states are capable of meeting the terms set out in these directives. Being very comprehensive, EU regulations include the following environmental issues:

- Water
- Air Pollution Control
- Waste
- Dangerous Substances; Chemicals
- Noise

### A. IPPC: Integrated Pollution Prevention and Control Directive

In 1996, the European Council published IPPC-Integrated Pollution Prevention and Control Directives, in order to regulate the permission and control procedures of industrial enterprises. IPPC forms the basis of the European Union Industrial Regulations on environmental issues. This is because these directives replaced the earlier EU regulations that were based on the receiving environment. They offer a comprehensive permission procedure that evaluates all the receiving environments together. An individual permission procedure applies for each receiving environment (ÇOB, 2006: 33).

IPPC has 4 fundamentals: best available techniques, emission limit values, people’s access to knowledge, and sharing knowledge. These points are summarised below.

The BAT (Best Available Techniques) concept aims to prevent or decrease pollution of air, water and land, and thus aims to attain a high level of protection.

Industrial and agricultural enterprises make use of BREF (Best Available Techniques Reference Document) in order to improve their environmental status, via Best Available Techniques. These documents do not have any legally binding effects.

## B. Water Directives

- Council Directive 75/440/EEC was published under the title "Improvement of Water Quality," and concerned "the quality required of surface water intended for the abstraction of drinking water in the member states."
- Council Directive 79/869/EEC, 9 October 1979, concerned the methods of measurement and the frequency of sampling and analysis of surface water intended for the abstraction of drinking water in the Member States.
- Council Directive 91/271/EEC, 21 May 1991, concerned Urban Waste-water Treatment
- Council Directive 2000/60/EC, 23 October 2000, established a framework for the Community action in the field of water policy or, in short, the EU Water Framework Directive. This aimed to regulate policies on surface waters, floods, coastal waters and underground waters, so as to protect and improve water areas.

All these directives have been adopted. In addition, "Water Pollution Control Regulation" entered into force on 31.12.2004 (Official Gazette, 2004, Issue: 25687).

## C. Air Quality Management

In order to prepare the Air Framework Bill, to harmonise The Large Combustion Plant Directive (2001/80/EC) and the Air Quality Framework Directive (96/62/EC), a joint project that aimed to prepare legal regulations to strengthen the institutional structure of the Ministry of Environment was carried out with the German Federal Ministry for the Environment, Nature Conservation and Nuclear Safety. Within this project (TÜSİAD, 2007: 53):

- The Large Combustion Plant Directive (2001/80/EC) and the Air Quality Framework Directive (96/62/EC) were transferred into Turkish regulations.
- The National Air Quality Control Framework Law (Air Quality and Emission) Draft was prepared.
- Capacity was raised for Air Quality Management.
- Strategic action plans for the application of directives were developed.

## D. Waste Management Directives

Including 22 articles and entering into force on 17 May 2006, the Waste Framework Directive has attachments listing waste categories, disposal methods, and recycling procedures. The directive includes definitions of waste, and also covers waste categories. Disposal and recycling activities are also listed. The directives on waste that are supposed to be harmonised with EU Regulations are as follows:

- **Dangerous Waste Directive;** "Regulation on Dangerous Substance Control", 14 March 2005 (Official Gazette Issue 25755) is responsible for the whole procedure from the production of dangerous substances to their disposal.
- **SEVESO II (96/82/EC);** The Council Directive 96/82/EC, 9 December 1996 concerns the control of major-accident hazards involving dangerous substances and limits their consequences for man and the environment. A new draft for regulations was prepared. The directive is applicable to certain big establishments that contain dangerous substances (ÇOB, 2006: 34).
- **Packaging and Package Waste Directive;** "Renewed Regulation on Control of Packaging Waste", 06.11.2008 (Official Gazette Issue 27046) covers all domestically produced plastic, metal, glass, paper-carton, composite and other packages and their wastes, without regard to their being domestic, industrial, commercial, or office waste. The regulation includes these objectives (Official Gazette, 2008, Issue: 27046);
- **Directive on Waste Oil Disposal;** "Regulation on Waste Oil Disposal", 21.01.2004 (Official Gazette Issue 25353) has entered into force. The regulation includes prohibitions, limitations, responsibilities, precautions, investigations, and legal and criminal liabilities with regard to the production, temporary storage, gathering, transport, recycling, and disposal, trade, import and export and transit pass of the waste oil specified in categories I, II, and III (Official Gazette, 2004, Issue: 25353).
- **Directive on Batteries and Accumulators with Dangerous Substances;** "Regulation on Control of Waste Batteries and Accumulators", 31 August 2004 (Official Gazette Issue 25569) is responsible for the whole procedure, from the production of batteries and accumulators to their final disposal (Official Gazette, 2004).

- **Waste Framework Directive;** "Regulation on Waste Management", 05.07.2008 (Official Gazette Issue 26927) aims to identify the main principles concerning the management of wastes from their creation to their final disposal, without damage to public health or the environment. The regulation covers the wastes listed in its supplement (Official Gazette, 2008, Issue: 26927).
- **Landfill of Waste Directive;** This directive (1999/31/EC) is intended to prevent or reduce the adverse effects of the landfill of waste on the environment, in particular its effects on surface water, groundwater, soil, air, and human health.
- **Incineration of Waste Directive;** Incineration of Waste Directive (2000/76/EC) includes technical clauses about the incineration of waste. This regulation, which entered into force on 06.10.2010, includes minimum requirements for incineration or co-incineration of waste (Official Gazette 2010, Issue: 27721).
- **PCB/PCT Directive;** "Regulation on Control of Polychlorinated Biphenyls and Polychlorinated Terphenyls" entered into force on 27 December 2007 (Official Gazette Issue 26739). It sets out administrative and technical methods and principals concerning the disposal of Polychlorinated Biphenyls (PCB) and used substances and equipment (including Polychlorinated Biphenyls) without threatening environment and public health (Official Gazette 2007, Issue: 26739).
- **Directive That Forbids the Use of Certain Dangerous Substances in Electrical and Electronics Equipment;** Regulation on Control and Management of Waste Electrical and Electronics Goods covers all big and small home appliances, computers, communication, consumer and lighting equipment, electrical and electronic tools (excluding big industrial tools), toys, hobby and sports equipment, medical equipment (excluding infected products), tracking and control devices and automatic dispensers (Official Gazette, 2006, Issue: 26236).

## E. Directive on Dangerous Substances and Chemicals

"Regulation on Classification, Packaging and Labelling of Dangerous Substances and Concoctions", 26 December 2008 (Official Gazette, Issue 27092) aims to regulate the technical and managerial principles concerning the classification, packaging and labelling of dangerous substances and concoctions, in order to control and supervise these substances and chemicals and to prevent possible damage to public health and environment (Official Gazette, 2008, Issue: 27092).

### F. Noise Directive

"Regulation on Assessment and Management of Environmental Noise", 7 March 2008 (Official Gazette Issue 26809) aims to take necessary precautions against environmental noise and to maintain people's psychological and physical health and peace. The regulation covers the principles and criteria concerning the environmental noises that people are exposed to in areas of dense population, parks or other silent places in cities, silent places in open areas, schools, hospitals and other places vulnerable to noise and damage on structures caused by environmental vibrations (Official Gazette, 2008, Issue 26809).

## 3. Harmonisation of Turkish Environmental Legislation with EU

The most recent and significant improvements to Turkey's responsibilities for harmonisation with EU environmental legislations began in November 2000. According to the Accession Partnership Document, revised in April 2003, Turkey's responsibilities for harmonisation in the short and long term are as follows:

**Short Term**

- Putting into place and applying the Environmental Effect Assessment Directive.
- Creating a financial plan for harmonisation.
- Efforts towards reflection and application of the legislative *acquis communitaire* on issues like framework legislation, environment protection, water quality, integrated pollution prevention and control, and waste management.

**Middle Term**

- Completing the *acquis communitaire* reflection.
- Improving the constitutional, administrative and tracking capacity to protect the environment.

A National Programme was created on 25 July 2003, to undertake the EU *Acquis Communitaire* by Turkey. The priorities of the National Programme are as follows:

- Improving Water Quality.
- Waste Management.
- Improving Air Quality.
- Nuclear Safety.
- Genetically Modified Organisms.
- Environmental Noise Management.
- Management of Chemicals.
- Strengthening the Environmental Effect Assessment process.

## 4. Survey

SMEs are located in every region of Turkey, and they play a major role in regional development. Both in great industrial zones and small industries, SMEs are of both economic and social benefit to their environments. 37.7% of added value in Turkey is produced by SMEs.

It is quite obvious that enterprises, especially SMEs and industrial zones, have a great influence on our society. This survey focuses on manufacturing enterprises, both inside and outside industrial zones, so as to assess possible hazards caused by enterprises' use of natural resources, to take environmental protection precautions, to assess the enterprises' level of responsibility for environmental issues, and to identify what needs to be done to ensure continuity. The reasons for this choice are as follows:

- Enterprises use environmental resources more than any other units.
- During their activities, some environmental pollution effects occur.
- During and after their activities, they produce waste that may affect public health.
- They have huge environmental responsibilities within society.
- SMEs and IZs make a major contribution to and have a great influence on society's development.
- Enterprises are among the priority groups identified by EU Directives.

### A. Objective of the Study

The study aims to find out to what extend SMEs, both inside and outside industrial zones, are knowledgeable about EU environmental directives and Turkish environmental legislation. The topics listed below have also been tested via our analyses:

1. To find out whether enterprises inside and outside the IZs differ in terms of their knowledge about EU environmental directives.
2. To find out whether enterprises with and without an environmental policy differ in terms of their knowledge about EU environmental directives.
3. To find out whether enterprises with and without a waste management unit differ in terms of their knowledge about EU environmental directives.
4. To find out whether enterprises with a small or a large number of employees differ in terms of their knowledge about EU environmental directives.

## B. Survey Methodology

The survey population covers manufacturing enterprises with more than 10 employees in the city of Sakarya. Enterprises were chosen by means of random sampling of the data from the Sakarya Chamber of Commerce and Industry. The questionnaire survey was applied from 1 January 2009 to 31 March 2009. Questionnaires were completed face to face with firm managers or owners and accountants.

Of 89 firms operating in the industrial zone of Sakarya, 60 were covered in the survey. 67.42 % of questionnaire feedback was received from the industrial zones. Of 110 firms outside the industrial zone, 56 responded to the questionnaire, with 50.9 % questionnaire feedback. In total, the questionnaires of 112 firms out of a total of 199 were taken into account, with 56.28 % questionnaire feedback. The SPSS 16.0 for Windows package programme was employed for the purpose of evaluating the questionnaire data.

## C. Survey Results

As the purpose of the survey was to compare firms inside and outside industrial zones in Sakarya, firms outside IZs were chosen according to the sectorial distribution of firms inside IZs. 58 enterprises inside IZs and 54 outside IZs participated in our questionnaire. The greatest participation came from firms in the vehicle transportation and supply industry. In the Sakarya Chamber of Commerce and Industry's Industrial Guide (and on their web site), sectors are divided into 10 groups. Since the $10^{th}$ group was categorised as "other firms," no such group was included in the survey.

Via the “Regulation on Definition, Qualities and Classification of Small and Medium Sized Enterprises” of 18.11.2006 (Official Gazette, Issue 997), harmonisation with the EU—in terms of employment—was achieved. According to this document, enterprises with less than 10 employees are micro enterprises, enterprises with less than 50 employees are small enterprises, and enterprises with less than 250 employees are medium-sized enterprises (SATSO, 2008: 86). Micro enterprises were not included in our study.

As the data in Table 2-3 shows, mostly small-sized enterprises work in IZs. Of 14 enterprises with more than 250 employees, 4 are inside IZs and 10 outside. In total, of 53 enterprises with 50-249 employees, 36 are inside IZs and 17 outside. Enterprises in IZs are mostly small-sized enterprises with 50-249 employees, and 18 enterprises have between 10-49 employees. Considering the number of employees, 27 of the enterprises outside IZs that participated in our survey are small-sized, 17 are medium-sized and 10 are big. As regards the time of establishment of the enterprises within the survey, the largest group inside IZs are 1-10 years old (24.1 %), and the largest outside IZs are 11-20 years old (22.3 %). This is possibly because the first IZ in the city was established in 1993, the second in 1997, and the third after 1999.

As Table 2-4 reveals, out of 48 enterprises in the survey that claimed to have a written and legal environmental policy, 27 are inside IZs and 21 outside. Out of 64 enterprises that do not have a written environmental policy, 31 are inside IZs and 33 outside. 48% of enterprises claimed to have an environmental policy, and 57.1% did not. The number of enterprises with a written environmental policy in IZs is greater than the number outside, as can be seen from Table 2-4.

As can be observed from Table 2-5, of 26 enterprises that claimed to have a waste management unit, 13 are inside IZs and 13 outside. Out of 86 enterprises that do not have a waste management unit, 45 are inside IZs and 42 outside.

Enterprises were questioned about 8 of the EU directives. Enterprises inside and outside IZs were evaluated separately. As Table 2-6 reveals, 45 out of 54 enterprises outside IZs are not knowledgeable about IPPC-Integrated Pollution Prevention and Control Directive, while 6 are. As regards BREF (Best Available Techniques Reference) Document, 46 enterprises are not knowledgeable, while 4 are. On SEVESO Directive, 45 enterprises are not knowledgeable, 6 are knowledgeable and 1 is very knowledgeable.

## Table 2-3 Sectorial Distribution of Enterprises in Sakarya

| Sectors | | Outside IZ | | Inside IZ | | Total | (%) |
|---|---|---|---|---|---|---|---|
| | | Freq. | (%) | Freg. | (%) | | |
| Food | | 11 | 9.8 | 8 | 7.1 | 19 | 17 |
| Machinery | | 3 | 2.7 | 7 | 6.2 | 10 | 8.9 |
| Electrical | | 2 | 1.8 | 2 | 1.8 | 4 | 3.6 |
| Transport Vehicles-Car Supplier Ind. | | 11 | 9.8 | 12 | 10.7 | 23 | 20.5 |
| Metal-Steel-Iron | | 4 | 3.6 | 9 | 8 | 13 | 11.6 |
| Construction | | 8 | 7.1 | 6 | 5.4 | 14 | 12.5 |
| Textile-Hosiery-Shoes | | 3 | 2.7 | 8 | 7.1 | 11 | 9.8 |
| Wood Products-Furniture | | 8 | 7.1 | 2 | 1.8 | 10 | 8.9 |
| Patrol-Chemistry | | 4 | 3.6 | 4 | 3.6 | 8 | 7.1 |
| **Total** | | **54** | **48.2** | **58** | **51.8** | **112** | **100** |
| **Number of Employees** | | | | | | | |
| Small | 10- 49 | 27 | 24.1 | 18 | 16.1 | 45 | 40.2 |
| Medium | 50-250 | 17 | 15.2 | 36 | 32.1 | 53 | 47.3 |
| Big | >250 | 10 | 8.9 | 4 | 3.6 | 14 | 12.5 |
| **Total** | | **54** | **48.2** | **58** | **51.8** | **112** | **100** |
| **Date of Establishment** | | | | | | | |
| 2009-1999 | | 10 | 8.9 | 27 | 24.1 | 37 | 33 |
| 1998-1988 | | 25 | 22.3 | 20 | 17.9 | 45 | 40.2 |
| 1987-1977 | | 8 | 7.1 | 4 | 3.6 | 12 | 10.7 |
| <1976 | | 11 | 9.8 | 7 | 6.2 | 18 | 16.1 |
| **Total** | | **54** | **48.2** | **58** | **51.8** | **112** | **100** |

## Table 2-4 Availability of Environmental Policies in Enterprises

| Availability of Written Environmental Policy | Outside IZ | | Inside IZ | | Total | |
|---|---|---|---|---|---|---|
| | Frequency | (%) | Frequency | (%) | Frequency | (%) |
| No | 33 | 29.5 | 31 | 27.7 | 64 | 57.1 |
| Yes | 21 | 18.8 | 27 | 24.1 | 48 | 42.9 |
| **Total** | **54** | **48.2** | **58** | **51.8** | **112** | **100** |

**Table 2-5 Availability of Waste Management Unit in Enterprises**

| Availability of Waste Management Unit | Outside IZ | | Inside IZ | | Total | |
|---|---|---|---|---|---|---|
| | Frequency | (%) | Frequency | (%) | Frequency | (%) |
| No | 41 | 36.6 | 45 | 40.2 | 86 | 76.8 |
| Yes | 13 | 11.6 | 13 | 11.6 | 26 | 23.2 |
| **Total** | **54** | **48.2** | **58** | **51.8** | **112** | **100** |

On Water Quality Directives, 28 enterprises are not knowledgeable, 13 are knowledgeable and 4 are very knowledgeable. On Air Quality Directives, 24 enterprises are not knowledgeable, 12 are knowledgeable and 5 are very knowledgeable. On Waste Management Directives, 22 enterprises are not knowledgeable, 15 are knowledgeable and 4 are very knowledgeable. On Landfill of Waste Directive, 21 enterprises are not knowledgeable, 16 are knowledgeable and 4 are very knowledgeable. Finally, on Noise Pollution Directives, 19 enterprises are not knowledgeable, 15 are knowledgeable and 6 are very knowledgeable. As Table 2-6 reveals in detail, enterprises outside IZ are not very knowledgeable about EU directives concerning waste.

**Table 2-6 Level of Knowledge about EU Directives and Documents in Enterprises Outside Industrial Zones**

| EU Directives and Documents | Not Knowledgeable | | Not Enough Knowledgeable | | Knowledgeable | | Very Knowledgeable | | T |
|---|---|---|---|---|---|---|---|---|---|
| | F | (%) | F | (%) | F | (%) | F | (%) | F |
| IPPC Directive | 45 | 40.2 | 3 | 2.7 | 6 | 5.4 | 0 | 0 | 54 |
| BREF Document | **46** | 41.1 | 4 | 3.6 | 4 | 3.6 | 0 | 0 | 54 |
| SEVESO Directive | 45 | 40.2 | 2 | 1.8 | 6 | 5.4 | 1 | 0.9 | 54 |
| Water Quality Directives | 28 | 25 | 9 | 8 | 13 | 11.6 | 4 | 3.6 | 54 |
| Air Quality Directives | 24 | 21.4 | 13 | 11.6 | 12 | 10.7 | 5 | 4.5 | 54 |
| Waste Management Directives | 22 | 19.6 | 13 | 11.6 | 15 | 13.4 | 4 | 3.6 | 54 |
| Landfill of Waste Directives | 21 | 18.8 | 13 | 11.6 | **16** | 14.3 | 4 | 3.6 | 54 |
| Noise Pollution Directives | 19 | 17 | 14 | 12.5 | 15 | 13.4 | 6 | 5.4 | 54 |

**Table 2-7 Level of Knowledge about EU Directives and Documents in Enterprises Inside Industrial Zones**

| EU Directives and Documents | Not Knowledgeable | | Not Enough Knowledgeable | | Knowledgeable | | Very Knowledgeable | | T |
|---|---|---|---|---|---|---|---|---|---|
| | F | (%) | F | (%) | F | (%) | F | (%) | F |
| IPPC Directive | 51 | 45.5 | 7 | 6.2 | 0 | 0 | 0 | 0 | 58 |
| BREF Document | **52** | 46.4 | 6 | 5.4 | 0 | 0 | 0 | 0 | 58 |
| SEVESO Directive | 49 | 43.8 | 6 | 5.4 | 2 | 1.8 | 1 | 0.9 | 58 |
| Water Quality Directives | 28 | 25 | 17 | 15.2 | 11 | 9.8 | 2 | 1.8 | 58 |
| Air Quality Directives | 24 | 21.4 | 11 | 9.8 | 21 | 18.8 | 2 | 1.8 | 58 |
| Waste Management Directives | 22 | 19.6 | 9 | 8 | **25** | 22.3 | 2 | 1.8 | 58 |
| Landfill of Waste Directives | 25 | 22.3 | 12 | 10.7 | 19 | 17 | 2 | 1.8 | 58 |
| Noise Pollution Directives | 22 | 19.6 | 11 | 9.8 | **25** | 22.3 | 0 | 0 | 58 |

As Table 2-7 indicates, of 58 enterprises inside IZs, 51 are not knowledgeable at all about IPPC Directive, while 7 are not knowledgeable enough. As regards the BREF Document, 52 enterprises are not knowledgeable at all, and 6 are not knowledgeable enough. Concerning the SEVESO Directive, 49 enterprises are not knowledgeable at all, 2 are knowledgeable and 1 is very knowledgeable. On Water Quality Directives, 28 enterprises are not knowledgeable at all, 6 are knowledgeable, and 2 are very knowledgeable. On Air Quality Directives, 24 enterprises are not knowledgeable at all, 21 are knowledgeable, and 2 are very knowledgeable. On Waste Management Directives, 22 enterprises are not knowledgeable at all, 25 are knowledgeable, and 2 are very knowledgeable. On Landfill of Waste Directives, 25 enterprises are not knowledgeable at all, 19 are knowledgeable, and 2 are very knowledgeable. Finally, on Noise Pollution Directives, 22 enterprises are not knowledgeable at all, while 25 are knowledgeable.

Enterprises inside IZs are not very knowledgeable about EU directives concerning waste. In the questionnaire, enterprises inside IZs ticked the option "not knowledgeable at all" for all directives more often than enterprises outside IZs. The number of enterprises that ticked "knowledgeable" was greater aming enterprises inside IZs than those outside. The choice of "Very knowledgeable" was almost three times more frequent among enterprises outside IZs than among those inside.

## D. Analysis

The hypotheses tested in the survey are:

**H1:** Enterprises with and without an environmental policy have a different level of knowledge about European Union environmental directives.

**H2:** Enterprises with and without a waste management unit have a different level of knowledge about European Union environmental directives.

**H3:** Enterprises inside and outside IZ have a different level of knowledge about European Union environmental directives.

**H4:** Big enterprises and small enterprises have a different level of knowledge about European Union environmental directives.

As shown in Table 2-8, a T-test (Independent Samples T-test) was used among independent groups, in order to test the relation between the availability of environmental policy and the level of knowledge about EU environmental directives in enterprises. According to the **Levene Test;** sig (2-tailed) for questions 3, 4, 5, 6, 7 and 8 is $\leq 0.05$ and a significant difference is observed. Thus, **H1** (Enterprises with and without an environmental policy have a different level of knowledge about European Union environmental directives) is confirmed for directives 3, 4, 5, 6, 7 and 8. For other directives, sig. (2-tailed) value is $>0.05$, and thus the hypothesis is denied.

**Table 2-8 Relation between Availability of Environmental Policy and Level of Knowledge about EU Environmental Directives**

| Environmental Policy and Directives | Answer | N | Mean | s.d. | t | d.f. | sig. (2 tailed) | P (sig.) |
|---|---|---|---|---|---|---|---|---|
| 1. IPPC Directive | Yes | 64 | .11 | .362 | -2.088 | 110 | .000 | .039 |
| | No | 48 | .31 | .657 | -1.933 | 68.125 | | .057 |
| 2. Bref Document | Yes | 64 | .11 | .362 | -1.381 | 110 | .006 | .170 |
| | No | 48 | .23 | .555 | -1.302 | 75.934 | | .197 |
| **3. SEVESO Directive** | Yes | 64 | .09 | .344 | -3.310 | 110 | .000 | .001 |
| | No | 48 | .50 | .899 | -2.972 | 57.345 | | **.004*** |
| **4. Water Quality Directives** | Yes | 64 | .55 | .815 | -3.726 | 110 | .086 | **.000*** |
| | No | 48 | 1.19 | 1.003 | -3.618 | 88.814 | | .000 |
| **5. Air Quality Directives** | Yes | 64 | .75 | .891 | -3.085 | 110 | .073 | **.003*** |
| | No | 48 | 1.31 | 1.035 | -3.020 | 92.471 | | .003 |
| **6. Waste Management Directives** | Yes | 64 | .86 | .941 | -2.708 | 110 | .681 | **.008*** |
| | No | 48 | 1.35 | .978 | -2.693 | 99.193 | | .008 |
| **7. Landfill of Waste Directive** | Yes | 64 | .77 | .904 | -3.180 | 110 | .389 | **.002*** |
| | No | 48 | 1.33 | .975 | -3.146 | 97.056 | | .002 |
| **8. Noise Pollution Directives** | Yes | 64 | .86 | .924 | -3.134 | 110 | .838 | **.002*** |
| | No | 48 | 1.42 | .942 | -3.125 | 100.328 | | .002 |

For Table 2-9, a T-test (Independent Samples T-test) was used among independent groups, in order to discover the relation between the availability of waste management units and the level of knowledge about EU environmental directives in enterprises. According to the **Levene Test;** sig (2-tailed) for directives 4.5 and 7 is ≤0.05, and a significant difference is observed. Thus, **H2** (Enterprises with and without a waste management unit have a different level of knowledge about European Union environmental directives) is confirmed for directives 4.5 and 7. For other directives, sig. (2-tailed) value is >0.05, and thus the hypothesis is denied. Significant differences are observed in the following case: 4 (Water Quality Directives), 5 (Air Quality Directives), and 7 (Landfill of Waste Directives).

**Table 2-9 Relation between Availability of Waste Management Unit and Level of Knowledge about EU Environmental Directives**

| Waste Management Unit and Directives | Answer | N | Mean | s.d. | t | d.f. | sig. (2 tailed) | P (sig.) |
|---|---|---|---|---|---|---|---|---|
| 1. IPPC Directive | Yes | 86 | .17 | .465 | -.818 | 110 | .074 | .415 |
| | No | 26 | .27 | .667 | -.677 | 32.686 | | .503 |
| 2. Bref Document | Yes | 86 | .16 | .457 | .087 | 110 | .884 | .931 |
| | No | 26 | .15 | .464 | .086 | 40.746 | | .932 |
| 3. SEVESO Directive | Yes | 86 | .21 | .576 | -1.694 | 110 | .001 | .093 |
| | No | 26 | .46 | .905 | -1.342 | 31.372 | | .189 |
| **4. Water Quality Directives** | Yes | 86 | .69 | .844 | -2.824 | 110 | .008 | .006 |
| | No | 26 | 1.27 | 1.151 | -2.396 | 33.526 | | **.022*** |
| **5. Air Quality Directives** | Yes | 86 | .88 | .926 | -2.118 | 110 | .094 | **.036*** |
| | No | 26 | 1.35 | 1.129 | -1.904 | 35.754 | | .065 |
| 6. Waste Management Directives | Yes | 86 | .98 | .945 | -1.873 | 110 | .402 | .064 |
| | No | 26 | 1.38 | 1.061 | -1.760 | 37.793 | | .086 |
| **7. Landfill of Waste Directive** | Yes | 86 | .85 | .901 | -3.307 | 110 | .391 | **.001*** |
| | No | 26 | 1.54 | 1.029 | -3.079 | 37.353 | | .004 |
| 8. Noise Pollution Directive | Yes | 86 | 1.01 | .927 | -1.738 | 110 | .254 | .085 |
| | No | 26 | 1.38 | 1.061 | -1.616 | 37.272 | | .115 |

For Table 2-10, a T-test (Independent Samples T-test) was used among independent groups, to test the relation between being located inside or outside IZs and the level of knowledge about EU environmental directives in enterprises. According to the **Levene Test;** sig (2-tailed) for all directives is >0.05. Thus, **H3** (Enterprises inside and outside IZs have a different level of knowledge about European Union environmental directives) is denied for all directives.

**Table 2-10 Relations between Being Inside or Outside IZ and Level of Knowledge about EU Environmental Directives**

| Being Inside or Outside IZ and Directives | Answer | N | Mean | s.d. | t | d.f. | sig. (2 tailed) | P (sig.) |
|---|---|---|---|---|---|---|---|---|
| 1. IPPC Directive | Inside IZ | 54 | .28 | .656 | 1.618 | 110 | .000 | .108 |
| | Outside IZ | 58 | .12 | .329 | 1.584 | 76.737 | | .117 |
| 2. Bref Document | Inside IZ | 54 | .22 | .572 | 1.382 | 110 | .004 | .170 |
| | Outside IZ | 58 | .10 | .307 | 1.355 | 79.939 | | .179 |
| 3. SEVESO Directive | Inside IZ | 54 | .31 | .748 | .713 | 110 | .119 | .477 |
| | Outside IZ | 58 | .22 | .594 | .707 | 101.073 | | .481 |
| 4. Water Quality Directives | Inside IZ | 54 | .87 | 1.029 | .524 | 110 | .083 | .602 |
| | Outside IZ | 58 | .78 | .879 | .521 | 104.623 | | .604 |
| 5. Air Quality Directives | Inside IZ | 54 | .96 | 1.027 | -.288 | 110 | .936 | .774 |
| | Outside IZ | 58 | 1.02 | .964 | -.288 | 108.023 | | .774 |
| 6. Waste Management Directives | Inside IZ | 54 | 1.02 | 1.000 | -.547 | 110 | .577 | .585 |
| | Outside IZ | 58 | 1.12 | .975 | -.547 | 108.964 | | .586 |
| 7. Landfill of Waste Directive | Inside IZ | 54 | 1.06 | .998 | .488 | 110 | .867 | .627 |
| | Outside IZ | 58 | .97 | .955 | .487 | 108.515 | | .627 |
| 8. Noise Pollution Directive | Inside IZ | 54 | 1.15 | 1.035 | .525 | 110 | .434 | .600 |
| | Outside IZ | 58 | 1.05 | .907 | .523 | 105.626 | | .602 |

For Table 2-11, a One-Way ANOVA technique was used among independent groups, to analyse the relation between the number of employees and the level of knowledge about EU environmental directives in enterprises. The result of the analysis is P≤0.05 for all directives, apart from the IPPC Directive. No difference concerning the IPPC Directive was observed among the groups. For other directives, there is a significant relation among the number-of-employees groups concerning the relation between the number of employees and the level of knowledge about EU Directives. Having analysed these results with a Scheffe Test, we can easily see which personnel sizes have a significant difference in terms of the directives (these are marked (*) on the table above).

- On SEVESO Directives, between 1-49 employees and > 250 employees, a difference was observed in favour of > 250 employees.
- On Water Quality Directives, between 1-49, 50-249 and > 250 employees, a difference was observed in favour of > 250 employees over both other groups.

- On Air Quality Directives, between 1-49, 50-249 and > 250 employees, a difference was observed in favour of > 250 employees over both other groups.
- On Waste Management Directives, between 1-49 and > 250 employees, a difference was observed in favour of > 250 employees; between 1-49 and 50-249 employees, a difference was observed in favour of 50-249 employees.
- On Landfill of Waste Directives, between 1-49, 50-249 and > 250 employees, a difference was observed in favour of > 250 employees over both other groups.
- On Noise Pollution Directives, between 1-49, 50-249 and > 250 employees, a difference was observed in favour of > 250 employees over both other groups.

A significant difference is observable between small and big enterprises, as regards the level of knowledge about EU Directives.

The Bref Document has P value $\leq 0.05$ in One-Way ANOVA, and $P>0.05$ in the Scheffe test. According to the Scheffe test results for six other directives, "**H4** (Big enterprises and small enterprises have a different level of knowledge about European Union environmental directives) is confirmed. For the other two directives with a result of $P>0.05$, the hypothesis is denied.

**Table 2-11 Relation between Personnel Size and EU Environmental Directives**

| Directives | F | P (sig.) | Personnel Size Groups | | P(sig.) |
|---|---|---|---|---|---|
| 1.IPPC Directive | .820 | .443 | | | |
| 2.Bref Document | 3.156 | **.047** | 1- 49 | 50-249 | .204 |
| | | | | >250 | .078 |
| | | | 50-249 | 1- 49 | .204 |
| | | | | >250 | .541 |
| | | | >250 | 1- 49 | .078 |
| | | | | 50-249 | .541 |
| 3.SEVESO Directive | 5.123 | **.007** | 1- 49 | 50-249 | .272 |
| | | | | >250 | **.008*** |
| | | | 50-249 | 1- 49 | .272 |
| | | | | >250 | .111 |
| | | | >250 | 1- 49 | **.008*** |
| | | | | 50-249 | .111 |
| 4.Water Quality Directives | 10.916 | **.000** | 1- 49 | 50-249 | .298 |
| | | | | >250 | **.000*** |
| | | | 50-249 | 1- 49 | .298 |
| | | | | >250 | **.002*** |
| | | | >250 | 1- 49 | **.000*** |
| | | | | 50-249 | **.002*** |
| 5.Air Quality Directives | 7.423 | **.001** | 1- 49 | 50-249 | .191 |
| | | | | >250 | **.001*** |
| | | | 50-249 | 1- 49 | .191 |
| | | | | >250 | **.033*** |
| | | | >250 | 1- 49 | **.001*** |
| | | | | 50-249 | **.033*** |
| 6.Waste Management Directives | 7.929 | **.001** | 1- 49 | 50-249 | **.019*** |
| | | | | >250 | **.002*** |
| | | | 50-249 | 1- 49 | **.019*** |
| | | | | >250 | .221 |
| | | | >250 | 1- 49 | **.002*** |
| | | | | 50-249 | .221 |
| 7.Landfill of Waste Directives | 6.814 | **.002** | 1- 49 | 50-249 | .913 |
| | | | | >250 | **.002*** |
| | | | 50-249 | 1- 49 | .913 |
| | | | | >250 | **.005*** |
| | | | >250 | 1- 49 | **.002*** |
| | | | | 50-249 | **.005*** |
| 8.Noise Pollution Directives | 5.471 | **.005** | 1- 49 | 50-249 | .858 |
| | | | | >250 | **.007*** |
| | | | 50-249 | 1- 49 | .858 |
| | | | | >250 | **.016*** |
| | | | >250 | 1- 49 | **.007*** |
| | | | | 50-249 | **.016*** |

## 5. Conclusion

The survey concludes that enterprises have very little knowledge about directives and documents of harmonisation with the European Union. Furthermore, the analyses reveal a significant difference between big and small enterprises concerning the level of knowledge about European Union Directives.

There is a significant difference between the availability of an environmental policy and the level of awareness about European Union Directives. The analysis also reveals a significant difference between the availability of a waste management unit and the level of awareness about European Union Directives. Enterprises with an environmental management unit and a waste management unit are more knowledgeable about European Union Directives.

Concerning the relation between personnel size and awareness of European Union Directives, a significant difference is observed, for all directives, in favour of enterprises with more than 250 employees. This proves that big enterprises have a greater awareness of European Union Directives than small enterprises.

There is no significant difference between enterprises inside and outside Industrial Zones concerning the awareness on European Union Environmental Directives. Industrial Zones, created as a necessity of planned development, are the best model for minimising the possible damage of industrialisation on the environment. Using this to our advantage, we must build an Environmental Management System and waste minimisation systems for every IZ. We must consider, in detail, issues such as what precautions must be takn and at what stage IZs must adopt this system. The system must be dynamic, must be revised every year, and changes in factories and IZs must be recorded.

Efforts towards harmonisation with EU Regulations are being carried out by the state and related parties. Nevertheless, harmonisation of environmental regulations seems to be one of the toughest issues for Turkey. This is mostly due to the high costs involved and the infrastructure needed to protect the environment and to eradicate current pollution.

One can say that big enterprises in Turkey have started following international developments on environment, and have also started taking precautions regarding this issue. Thanks to EU efforts, big enterprises with overseas connection that export goods have come to realise the importance of their responsibilities concerning the environment. It is these big enterprises that work to obtain certificates related to the environment, such as ISO 14001. SMEs, on the other hand, are not sensitive enough to environmental issues. The reasons for this are as follows: leading people in SMEs do not attach enough importance to this matter, SMEs don't have a large enough budget, SMEs do not believe in the benefits of these efforts, and SMEs lack the required technology.

The fact that there are still some enterprises that have no idea about which directive covers their product proves that general directives prepared by the EU must be presented to producer enterprises in Turkey with more explanatory information and footnotes. Chambers of industry and the Small and Medium Enterprises Development Organisation are supposed to be the main drivers of activity on this issue.

IZ managements must ensure the participation of factories in the environment management system to be established, and programmes must be applicable and accepted by majority. IZ managements must start Environment Management Systems for enterprises under their control, and support environmental sensitivity in firms (especially SMEs). Companies must change their way of thinking about environmental activities and accept the fact that environment is a major factor in the development of business.

## References

ÇOB, (Çevre ve Orman Bakanlığı), (2007), Çevre Denetçileri El Kitabı, Ankara http://www.cevredenetimi.cevreorman.gov.tr/DOSYALAR/denetim-el-kitabi.pdf, 04.07.2009.

ÇOB, (Çevre ve Orman Bakanlığı), (2006), "Sakarya İl Çevre Durum Raporu", http://www.cedgm.gov.tr/icd_raporlari/sakaryaicd2007.pdf s: 326, 29.06.2009.

ÇOB, (Çevre ve Orman Bakanlığı), Çevre Çalışmaları ve Yatırımları (2003-2008), http://www.cevreorman.gov.tr/belgeler6/cevre0308.pdf, 30.06.2009.

Kelgökmen, Derya, (2006), Avrupa Birliği'nde Çevre Koruma Politikaları ve Türkiye'deki Yerli ve Yabancı Firmaların Pazarlama Politikalarına Etkileri, Basılmamış Yüksek Lisans Tezi, Ege Üniversitesi Sosyal Bilimler Enstitüsü.

Resmi Gazete, (2008), "Atıkların Yakılmasına İlişkin Yönetmelik", Tarih: 06.10.2010, Sayısı: 27721, http://www.cygm.gov.tr/CYGM/Files/mevzuat/yonetmelik/ATIKLARIN%20YAKILMASINA%20İLİŞKİN%20YÖNETMELİK.doc.

—. (2008), Tehlikeli Maddelerin ve Müstahzarların Sınıflandırılması, Ambalajlanması ve Etiketlenmesi Hakkında Yönetmelik, Tarih: 26.12.2008, Sayı: 27092-Mükerrer, http://www.mevzuat.adalet.gov.tr/html/28139.html, 30.06.2009.

—. (2008), Çevresel Gürültünün Değerlendirilmesi Ve Yönetimi Yönetmeliği, Tarih: 7 Mart 2008, Sayı: 26809, http://rega.basbakanlik.gov.tr/eskiler/2008/03/20080307-6.htm, 30.06.2009.

—. (2008), Atık Yönetimi Genel Esaslarına İlişkin Yönetmelik, Çevre ve Orman Bakanlığı, Tarih: 05.07.2008, Sayı: 26927, http://www.atikyonetimi.cevreorman.gov.tr/yonetmelikler/Genelesas.doc, 29.06.2009.

—. (2008), Ambalaj Atıklarının Kontrolü Yönetmeliği, Tarih: 06.11.2008, Sayı: 27046, http://www.atikyonetimi.cevreorman.gov.tr/yonetmelikler/ambalaj.doc, 29.06.2009.

—. (2007), Poliklorlu Bifenil Ve Poliklorlu Terfenillerin Kontrolü Hakkında Yönetmelik, Tarih: 27 Aralık 2007, Sayı: 26739, http://rega.basbakanlik.gov.tr/eskiler/2007/12/20071227-3.htm, 04.07.2009.

—. (2006), Endüstri Tesislerinden Kaynaklanan Hava Kirliliğinin Kontrolü Yönetmeliği, Tarih: 22 Temmuz 2006, Sayı: 26236.

—. (2005), Tehlikeli Atıkların Kontrolü Yönetmeliği, Tarih: Tarih: 14 Mart 2005, Sayı: 25755, http://www.atikyonetimi.cevreorman.gov.tr/yonetmelikler/ tehlike.doc, 29.06.2009.

—. (2004), Su Kirliliği Kontrolü Yönetmeliği, Tarih: 31 Aralık 2004, Sayı: 25687, http://www.cevreorman.gov.tr/yasa/y/25687.doc, 30.06.2009.

—. (2004), Atık Yağların Kontrolü Yönetmeliği, Tarih: 21 Ocak 2004, Sayı: 25353, http://www.cevreorman.gov.tr/yasa/y/25353.doc, 30.06.2009.

—. (2004), Atık Pil ve Akümülatörlerin Kontrolü Yönetmeliği, Tarih: 31 Ağustos 2004, Sayı: 25569.

SATSO, (Sakarya Ticaret ve Sanayi Odası), (2008), 2008 Yılı İktisadi Raporu, hazırlayanlar: Uğur TIN ve diğ., Merkez Matbaacılık, Sakarya.

TÜSİAD, (Türk Sanayicileri ve İşadamları Deneği), (2007), Sanayide AB Çevre Mevzuatına Uyum, Yay.No: TÜSİAD-T/2007-05-440, İstanbul.

# WIND SPEED AND ENERGY CHARACTERISTICS BASED ON STATISTICAL ANALYSES

ZAFER ASLAN[1] AND Z. NEVIN ÇAĞLAR[2]

## 1. Introduction

Due to the development of wind energy technologies, most private companies and governmental organizations have, since the 1990s, become interested in renewable energy sources (Tolun et al, 1995; Altunc et al., 2002; Aslan and Gençoğlu, 2005; Tokgozlu et al., 2010; Akdağ and Güler, 2008; Akkdağ et al., 2010). Turkey is known to have dynamic economic development and rapid population growth. As Turkey's economy experienced high levels of growth in the mid-1990s, the country's industrial production resulted in high levels of pollution and greater risks to the country's environment. Turkish energy consumption has risen dramatically over the past 20 years. From just 24 Mtoe in 1980, Turkey's domestic energy consumption has increased almost threefold, reaching a level of 77.1 Mtoe in 2000 (Kenisarin et al., 2006; Özdamar, 1999; Turksoy, 1995). Remote Sensing Data is also a useful tool for the purpose of defining the surface characteristics effect on temporal and spatial variations of wind velocity (Almalı and Ete, 2009; Aydın, 2009).

New and renewable energy markets are environmental friendly (Fells, 2000). Windmills have been successful in pumping water by mechanical energy from wells for livestock and homesteads thoughout the US and European countries for many years. The wind turbine drives the pump at varying speeds. When the wind speed reaches more than 4.5m/s, it can provide the same amount or more water than solar panels, diesel generators or old fashioned, mechanical windmills. This paper presents an overview of statistics, probability and wavelet analyses of wind speeds in

[1] Professor of Computer Engineering, Faculty of Engineering and Architecture, Istanbul Aydin University, TURKEY, zaferaslan@aydin.edu.tr
[2] Engineer, Meteorology Laboratory, Kandilli Observatory and Earthquake Research Institute, Bosphorus University, TURKEY, caglarz@boun.edu.tr

Şile (Istanbul), with a view to the implementation of wind energy, and in order to evaluate the results of previous analyses.

## 2. Description of Study Area and Data

### A. Study Area

The study area (Şile; latitude: 41,17oN, longitude: 29.6o E, height, 52m. above msl.) is situated in the northwestern part of Turkey. Short-term climatic variation in the Black Sea region is significantly influenced by the operation of North Atlantic Oscillation, a term used to describe the climatic mechanisms resulting from the interaction between the north Atlantic and mid-latitude air masses. While the exact mechanisms causing the North Atlantic Oscillation remain unclear, it is thought that the climate conditions established in Western Europe mediate the heat and precipitation fluxes reaching Central Europe and Eurasia, and regulate the formation of winter cyclones that are largely responsible for regional precipitation inputs and that influence Mediterranean Sea Surface Temperatures (SSTs). The relative strength of these systems also limits the amount of cold air that arrives from northern regions during winter. Another influential factor is the regional topography, as depressions and storms systems arriving from the Mediterranean are funnelled through the low land around the Bosphorus, Pontic and Caucasus mountain ranges act as wave guides, limiting the speed and paths of cyclones passing through the region.

### B. Data and Software

The Analysis of wind speed is based on measurements and observations at Şile Governmental Meteorological Observatory, which is located at 10m above mean ground level. In the present study, daily, monthly and annual wind speed values recorded between 1975 and 2010 will be analysed statistically.

Hourly, daily, monthly and seasonal variations of wind velocity characteristics of the study area will be analysed using Easy Fit5.5 Professional Evaluation Version (StatAssist), MATLAB-Wave Menu, and SPSS packets.

## C. Methodology

### i. Wind Velocity and Power

The presentation of wind data makes use of the Weibull Distribution Function, as a tool to represent the frequency distribution of wind speed. The cumulative distribution function is obtained by integrating the Weibull Distribution, and is expressed mathematically as follows (Tolun et al., 1995; Nemes, C. and F. Munteanu, 2011):

$$F(u) = \exp\left[-\left(\frac{u}{\alpha}\right)^{\beta}\right] \tag{1}$$

F(u) gives the probability of the wind speed exceeding the value u, α(m/s) is the scale parameter (m/s), β is the shape parameter, and u is the wind speed (m/s).

The wind power density (W/m2) over a time interval is given as:

$$E = \left(\frac{\rho}{2}\right)(u_{av})^3 \tag{2}$$

In Eq.2, air density (ρ) may be taken as constant (ρ = 1.225kg/m3), with an error of less than a few percent, except for mountainous areas and interior stations (Tolun et al., 1995), where uav is the average wind speed over a time interval T.

At high wind speeds, the wind profile over flat and reasonably homogeneous terrain is well modelled using the logarithmic law. Wind speed data are extrapolated using the following power law:

$$V_r = V_a\left(\frac{Z_r}{Z_a}\right)^{\mu} \tag{3}$$

Vr and Va are the extrapolated and observed wind speeds, Za is the anemometer height, Zr is the extrapolated height, and μ is the power law component. μ is a function of atmospheric stability, and is called the wind shear exponent or Helmann Coefficient (Eq.3 and Table 2-12) (Akdağ ve Güler, 2008).

**Table 2-12 Variation of Helmann Coefficient at Different Areas**

| Characteristics of Site Area | Helmann Coefficient |
|---|---|
| Open Sea Surface | 0.14 |
| Open Field, Farmlands | 0.18 |
| Forests and City | 0.28 |
| Urbanised Area | 0.4 |

For the purpose of this study, observations were taken in coastal parts of Şile in the Black Sea Region. Instead of Helman coefficients, the average value of μ is accepted as 0.2 Vegetation cover, and urbanised area and land-sea interactions were considered. In detailed analyses, wind directions and surface roughness should be considered using further programs like WAsP (Wind Atlas Analysis and Application Program), (Tolun et al., 1995; Türksoy, 1995; Akdağ et al., 2010).

## ii. Wavelet Techniques

The present study is based on wavelet techniques and their applications to wind data. In this section, we give the basic definitions of wavelet transform. Wavelets are families of small waves generated from a single function f(t), which is called the mother wavelet. A sufficient condition for f(t) to qualify as a mother wavelet is given below (Meyer, 2000; Can et al., 2005, Siddiqi et al., 2002; Siddiqi et al., 2005).

$$\int_{-\infty}^{\infty} | f(t) |^2 \, dt < \infty \qquad 4(a)$$

The Fourier transform F of f(t) is defined as

$$F(w) = \int_{-\infty}^{\infty} f(t) e^{iwt} \, dt \qquad 4(b)$$

A function ψ(t) satisfying the following condition is called a continuous wavelet:

$$\int_{-\infty}^{\infty} |\psi(t)|^2 \, dt = 1 \qquad 5(a)$$

and

$$\int_{-\infty}^{\infty} |\psi(t)| \, dt = 0 \qquad 5(b)$$

Here, a is a scaling parameter, b is a location parameter, and ψa,b(u) is often called a continuous wavelet (or daughter wavelet), while ψ(u) is the mother wavelet. It may be observed that the scalogram can be represented either as a three-dimensional plot or as a two-dimensional grey scale image. As mentioned above, a, b are parameters representing the scaling factor and the location in time (Siddiqi et al., 2005). In the following sections, f(t) will be considered as daily and monthly average values of wind speed in Şile (Istanbul).

## 3. Analyses

### i. Statistical Analyses, Weibull Distribution

Figures 2-2 a and b show variations in the monthly wind speed and standard deviation in Şile. Higher values of mean wind speed and standard deviations are observed in autumn and winter. Table 2-13a and 2-13b give fitting results and goodness of fit for different functions for Şile.

The monthly variations of Weibull shape, scale parameters and probability density function are presented in Figure 2-3 a-c respectively. Maximum (5.23) and minimum (3.30) shape parameters are observed in November and June. Maximum and minimum values of scale parameters are defined in March (5.0m/s) and in October (1.29m/s) respectively. The highest values of mean monthly wind speed are recorded during the winter, spring and autumn seasons. The highest values are associated with the decrease in temperature during these seasons. Decreasing air temperature results in thermal convection and the increase in momentum at the upper layers. These cause an increase in the mean monthly wind speed (Shada, 2008).

Figure 2-2 Monthly Variation of Wind Speed (a) and Standard Deviation (b) in Şile

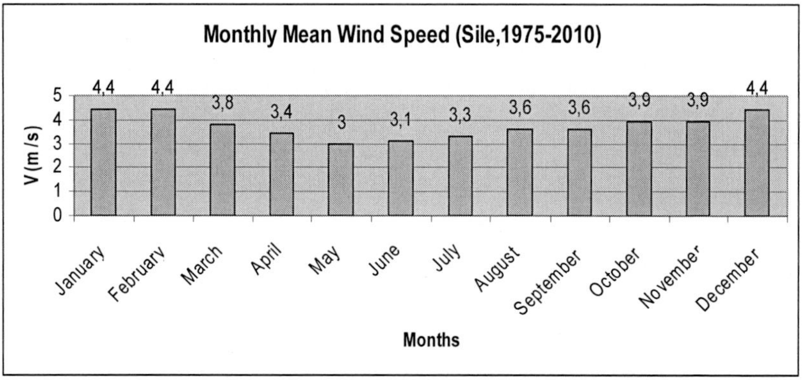


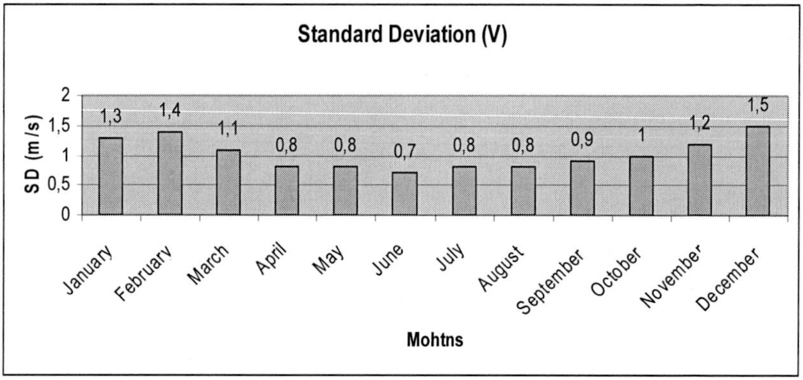

Figure 2-3 a-b show monthly variations of Weibull shape and scale parameters in Şile (1975-2010).

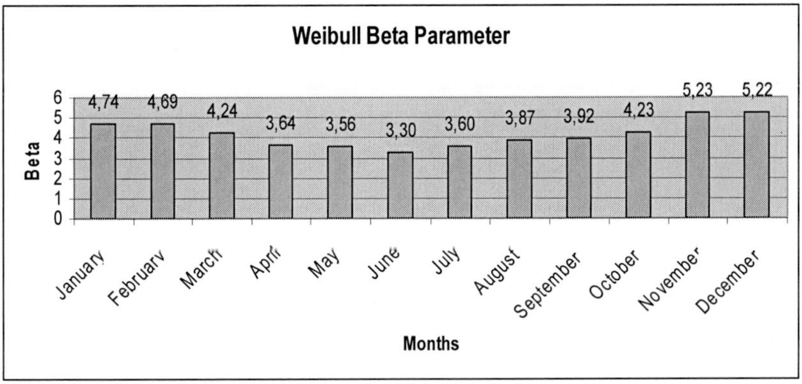


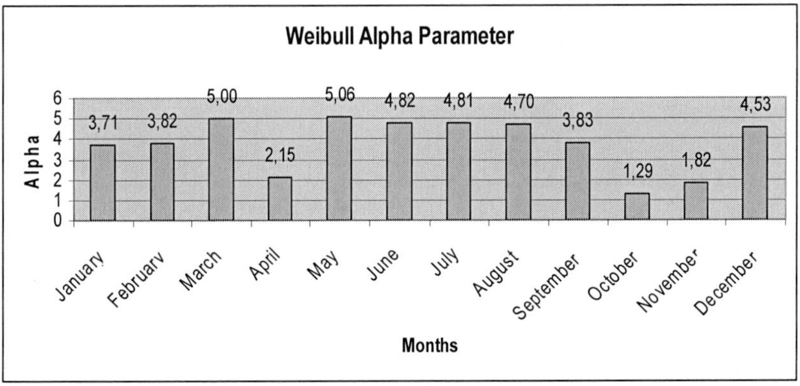

**Table 2-13 (a) Fitting Results of Different Distribution Functions of Long Term Monthly Wind Speed Values in Şile (1975-2010)**

| No | Distribution | Parameters | No | Distribution | Parameters |
|---|---|---|---|---|---|
| 1 | Beta | $□_1$=10.306 $□_2$=182.05 a=-0.7197 b=83.042 | 32 | Laplace | □=1.2339 □=3.7214 |
| 2 | Burr | k=2.4022 □=4.4874 □=4.6542 | 33 | Levy | □=3.072 |
| 3 | Burr | k=2.4022 □=4.4874 □=4.6542 | 34 | Levy | □=3.072 |
| 4 | Cauchy | □=0.66073 □=3.5965 | 35 | Log-Logistic | □=4.0957 □=3.5012 |
| 5 | Chi-Squared | □=4 | 36 | Log-Logistic | □=5.5579 □=3.6003 |
| 6 | Chi-Squared | □=3 | 37 | Logistic | □=0.63191 □=3.7214 |
| 7 | Dagum | k=0.53661 □=7.3088 □=4,1719 | 38 | Lognormal | □=0.37508 □=1.2615 |
| 8 | Dagum | k=0.53661 □=7.3088 □=4.1719 | 39 | Lognormal | □=0.37509 □=1.2615 |
| 9 | Erlang | m=9 □=0.40269 | 40 | Nakagami | m=2.5893 □=15.16 |
| 10 | Erlang | m=10 □=0.353 | 41 | Normal | □=1.1462 □=3.7214 |
| 11 | Error | k=1.3176 □=1.1462 □=3.7214 | 42 | Pareto 2 | □=352.66 □=1362.5 |
| 12 | Error Function | h=0.61694 | 43 | Pearson 5 | □=3.7518 □=11.526 |
| 13 | Exponential | □=0.26871 | 44 | Pearson 5 | □=3.7518 □=11.526 |
| 14 | Exponential (2P) | □=0.26871 □=-1.0000E-14 | 45 | Pearson 6 | $□_1$=10.273 $□_2$=2.7951E+5 □=1.0140E+5 |
| 15 | Fatigue Life | □=0.45238 □=3.5035 | 46 | Pearson 6 | $□_1$=9.3029 $□_2$=15009.0 □=6018.9 |

| | | | | | |
|---|---|---|---|---|---|
| 16 | Fatigue Life | □=0.45238 □=3.5035 | 47 | Pert | m=3.5518 a=-0.06153 b=8.4472 |
| 17 | Frechet | □=3.3673 □=3.109 | 48 | Power Function | □=1.1106 a=-1.0453E-5 b=8.3501 |
| 18 | Frechet | □=2.7031 □=2.8391 | 49 | Rayleigh | □=2.9693 |
| 19 | Gamma | □=9.2631 □=0.40269 | 50 | Rayleigh | □=2.7564 |
| 20 | Gamma | □=10.542 □=0.353 | 51 | Rice | □=3.531 □=1.1676 |
| 21 | Gen. Extreme Value | k=-0,13975□=1.0194 □=3.2582 | 52 | Student's t | □=8 |
| 22 | Gen. Gamma | k=0.97559 □=9.9508 □=0.353 | 53 | Triangular | m=3.32 a=-0.03674 b=8.3824 |
| 23 | Gen. Gamma | k=2.2948 □=2.1085 □=2.8512 | 54 | Uniform | a=1.7362 b=5.7066 |
| 24 | Gen. Pareto | k=-0.69278□=2.8749 □=2.0231 | 55 | Weibull | □=3.4156 □=4.1302 |
| 25 | Gumbel Max | □=0.89365 □=3.2056 | 56 | Weibull | □=2.9288 □=4.2501 |
| 26 | Gumbel Min | □=0.89365 □=4.2373 | 57 | Johnson SB | No fit |
| 27 | Hypersecant | □=1.1462 □=3.7214 | 58 | Log-Gamma | No fit |
| 28 | Inv. Gaussian | □=17.413 □=3.7301 | 59 | Log-Pearson 3 | No fit |
| 29 | Inv. Gaussian | □=39.233 □=3.7214 | 60 | Pareto | No fit |
| 30 | Johnson SU | □=-0.73797 □=2.3439 □=2.3199 □=2.9082 | 61 | Reciprocal | No fit |
| 31 | Kumaraswamy | □$_1$=2.036 □$_2$=4.3501 a=6.9147E-16 b=9.3329 | | | |

**Table 2-13 (b) Goodness of Fit–Summary**

| No | Distribution | Kolmogorov Smirnov | Anderson Darling | Chi-Squared | | | |
|---|---|---|---|---|---|---|---|
| | | Statistic | Rank | Statistic | Rank | Statistic | Rank |
| 1 | Beta | 0.06668 | 22 | 4.9181 | 21 | 31.222 | 28 |
| 2 | Burr | 0.029 | 2 | 2.6388 | 8 | 10.585 | 8 |
| 3 | Burr | 0.029 | 3 | 2.6388 | 9 | 10.585 | 7 |
| 4 | Cauchy | 0.10165 | 34 | 6.5012 | 26 | 65.998 | 32 |
| 5 | Chi-Squared | 0.2486 | 47 | 54.612 | 46 | 339.84 | 44 |
| 6 | Chi-Squared | 0.41603 | 52 | 119.63 | 52 | 474.38 | 45 |
| 7 | Dagum | 0.03312 | 8 | 2.7272 | 10 | 8.4626 | 4 |
| 8 | Dagum | 0.03312 | 9 | 2.7272 | 11 | 8.4626 | 5 |
| 9 | Erlang | 0.07714 | 27 | 6.4756 | 25 | 15.192 | 17 |
| 10 | Erlang | 0.09646 | 32 | 10.242 | 29 | 16.084 | 20 |
| 11 | Error | 0.07219 | 24 | 2.1658 | 6 | 17.129 | 22 |
| 12 | Error Function | 0.93625 | 55 | 2120.8 | 56 | 3350.6 | 53 |
| 13 | Exponential | 0.39627 | 50 | 100.24 | 50 | 654.56 | 48 |
| 14 | Exponential (2P) | 0.39627 | 51 | 98.089 | 49 | 654.56 | 49 |
| 15 | Fatigue Life | 0.11067 | 35 | 18.325 | 32 | 93.09 | 34 |
| 16 | Fatigue Life | 0.11067 | 36 | 18.325 | 33 | 93.09 | 35 |
| 17 | Frechet | 0.07783 | 28 | 19.996 | 35 | 39.558 | 29 |
| 18 | Frechet | 0.16687 | 44 | 38.203 | 43 | 112.63 | 37 |
| 19 | Gamma | 0.04374 | 14 | 3.9393 | 19 | 12.348 | 10 |
| 20 | Gamma | 0.03049 | 5 | 2.932 | 12 | 12.467 | 11 |
| 21 | Gen. Extreme Value | 0.02332 | 1 | 0.58533 | 2 | 6.9984 | 3 |
| 22 | Gen. Gamma | 0.04257 | 12 | 3.721 | 17 | 11.784 | 9 |
| 23 | Gen. Gamma | 0.03879 | 10 | 3.4725 | 15 | 10.467 | 6 |
| 24 | Gen. Pareto | 0.06438 | 21 | 52.823 | 45 | N/A | |
| 25 | Gumbel Max | 0.05483 | 17 | 2.0826 | 5 | 16.15 | 21 |

| 26 | Gumbel Min | 0.11581 | 37 | 20.265 | 36 | 57.574 | 31 |
|---|---|---|---|---|---|---|---|
| 27 | Hypersecant | 0.07078 | 23 | 2.2925 | 7 | 12.775 | 13 |
| 28 | Inv. Gaussian | 0.1394 | 39 | 21.682 | 37 | 96.1 | 36 |
| 29 | Inv. Gaussian | 0.04444 | 15 | 5.9902 | 24 | 15.731 | 18 |
| 30 | Johnson SU | 0.03174 | 6 | 0.48987 | 1 | 2.4185 | 1 |
| 31 | Kumaraswamy | 0.15823 | 40 | 27.43 | 39 | N/A | |
| 32 | Laplace | 0.09859 | 33 | 4.646 | 20 | 29.807 | 27 |
| 33 | Levy | 0.49936 | 54 | 140.14 | 53 | 1257.2 | 50 |
| 34 | Levy | 0.49936 | 53 | 140.14 | 54 | 1257.2 | 51 |
| 35 | Log-Logistic | 0.09031 | 29 | 12.051 | 31 | 49.296 | 30 |
| 36 | Log-Logistic | 0.03043 | 4 | 2.9527 | 13 | 12.636 | 12 |
| 37 | Logistic | 0.06012 | 20 | 1.6612 | 3 | 6.8737 | 2 |
| 38 | Lognormal | 0.07649 | 25 | 8.5673 | 27 | 27.394 | 25 |
| 39 | Lognormal | 0.0765 | 26 | 8.5687 | 28 | 27.395 | 26 |
| 40 | Nakagami | 0.04202 | 11 | 3.6324 | 16 | 15.762 | 19 |
| 41 | Normal | 0.04836 | 16 | 1.9724 | 4 | 13.228 | 16 |
| 42 | Pareto 2 | 0.38404 | 49 | 95.718 | 48 | 654.52 | 47 |
| 43 | Pearson 5 | 0.1637 | 43 | 34.479 | 42 | 174.61 | 42 |
| 44 | Pearson 5 | 0.1637 | 42 | 34.479 | 41 | 174.61 | 41 |
| 45 | Pearson 6 | 0.03196 | 7 | 3.0477 | 14 | 12.875 | 14 |
| 46 | Pearson 6 | 0.04312 | 13 | 3.8945 | 18 | 13.142 | 15 |
| 47 | Pert | 0.13235 | 38 | 19.289 | 34 | 117.8 | 38 |
| 48 | Power Function | 0.32291 | 48 | 74.527 | 47 | 498.93 | 46 |
| 49 | Rayleigh | 0.18787 | 45 | 32.923 | 40 | 191.58 | 43 |
| 50 | Rayleigh | 0.2235 | 46 | 39.035 | 44 | 161.06 | 40 |
| 51 | Rice | 0.05896 | 18 | 5.0546 | 22 | 18.702 | 23 |
| 52 | Student's t | 0.93649 | 56 | 1669.3 | 55 | 3350.2 | 52 |
| 53 | Triangular | 0.15826 | 41 | 22.902 | 38 | 131.31 | 39 |
| 54 | Uniform | 0.09073 | 30 | 100.33 | 51 | N/A | |

| | | | | | | | |
|---|---|---|---|---|---|---|---|
| 55 | Weibull | 0.05938 | 19 | 5.2491 | 23 | 19.374 | 24 |
| 56 | Weibull | 0.09344 | 31 | 11.345 | 30 | 67.683 | 33 |
| 57 | Johnson SB | No fit | | | | | |
| 58 | Log-Gamma | No fit | | | | | |
| 59 | Log-Pearson 3 | No fit | | | | | |
| 60 | Pareto | No fit | | | | | |
| 61 | Reciprocal | No fit | | | | | |

Figure 2-3 c- Probability Density Function of Monthly Mean Wind Speed Values (N = 430) in Şile (1975-2010)

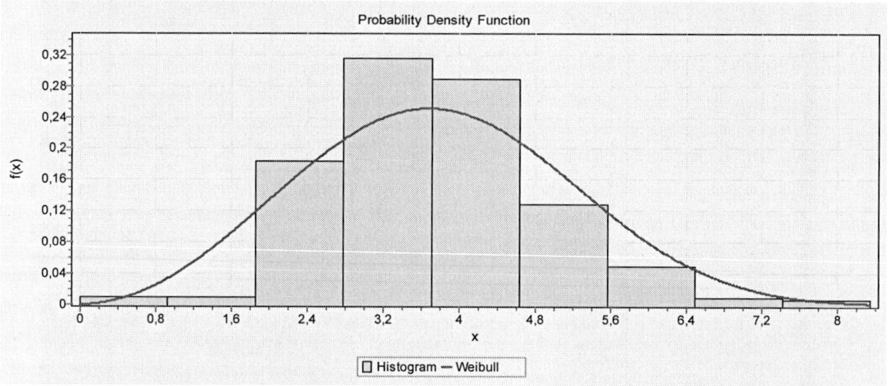


Table 2-14 gives an estimation of monthly mean wind speed values based on empirical formulae (Triola, 1998). With the values of 68%, 95% and 99.7%, probability, mean wind speed values change in the range between V±sd, V±2sd, and V±3sd respectively.

**Table 2-14 Estimation of Monthly Wind Speed in Şile based on the Empirical Approximation**

| Month | V10 | sd | V-sd | V+sd | V-2sd | V+2sd | V-3sd | V+3sd |
|---|---|---|---|---|---|---|---|---|
| January | 4.4 | 1.3 | 3.1 | 5.7 | 1.8 | 7 | 0.5 | 8.3 |
| February | 4.4 | 1.4 | 3 | 5.8 | 1.6 | 7.2 | 0.2 | 8.6 |
| March | 3.8 | 1.1 | 2.7 | 4.9 | 1.6 | 6 | 0.5 | 7.1 |
| April | 3.4 | 0.8 | 2.6 | 4.2 | 1.8 | 5 | 1 | 5.8 |
| May | 3.0 | 0.8 | 2.2 | 3.8 | 1.4 | 4.6 | 0.6 | 5.4 |
| June | 3.1 | 0.7 | 2.4 | 3.8 | 1.7 | 4.5 | 1 | 5.2 |
| July | 3.3 | 0.8 | 2.5 | 4.1 | 1.7 | 4.9 | 0.9 | 5.7 |
| August | 3.6 | 0.8 | 2.8 | 4.4 | 2 | 5.2 | 1.2 | 6 |
| September | 3.6 | 0.9 | 2.7 | 4.5 | 1.8 | 5.4 | 0.9 | 6.3 |
| October | 3.9 | 1.0 | 2.9 | 4.9 | 1.9 | 5.9 | 0.9 | 6.9 |
| November | 3.9 | 1.2 | 2.7 | 5.1 | 1.5 | 6.3 | 0.3 | 7.5 |
| December | 4.4 | 1.5 | 2.9 | 5.9 | 1.4 | 7.4 | -0.1 | 8.9 |
| Average | 3.733 | 1.025 | 2.708 | 4.758 | 1.683 | 5.783 | 0.658 | 6.808 |

Table 2-15 gives wind speed and wind power values defined by logarithmic law (Eq. 3) at 10, 30 and 50m above mean ground level (amgl) in Şile. Maximum values of wind power increase to 190,7kW/m2 in December.

**Table 2-15 Extrapolated Wind Speed and Available Wind Power (kW/m2month) at 10, 30 and 50m amgl in Şile**

| Month | 10m | E10 | 30m | E30 | 50m | E50 |
|---|---|---|---|---|---|---|
| January | 4.4 | 37.566 | 5.481 | 72.622 | 7.563 | 190.744 |
| February | 4.4 | 37.566 | 5.481 | 72.622 | 7.563 | 190.744 |
| March | 3.8 | 24.199 | 4.734 | 46.780 | 6.531 | 122.870 |
| April | 3.4 | 17.333 | 4.235 | 33.508 | 5.844 | 88.010 |
| May | 3 | 11.907 | 3.737 | 23.018 | 5.156 | 60.458 |
| June | 3.1 | 13.138 | 3.862 | 25.398 | 5.328 | 66.708 |
| July | 3.3 | 15.848 | 4.111 | 30.637 | 5.672 | 80.470 |
| August | 3.6 | 20.575 | 4.485 | 39.776 | 6.188 | 104.472 |
| September | 3.6 | 20.575 | 4.485 | 39.776 | 6.188 | 104.472 |
| October | 3.9 | 26.160 | 4.858 | 50.571 | 6.703 | 132.827 |
| November | 3.9 | 26.160 | 4.858 | 50.571 | 6.703 | 132.827 |
| December | 4.4 | 37.566 | 5.481 | 72.622 | 7.563 | 190.744 |
| Average | 3.73 | 22.947 | 4,651 | 44,361 | 6.417 | 116.515 |

## ii. Wavelet Analyses of Monthly Mean Wind Speed in Şile

In order to analyse monthly mean wind speed values recorded in Şile, MATLAB Wave-menu was applied. 1D Wavelet packet and 1D Continuous wavelet analyses explain the role of events on wind speed variation at different scales.

Figure 2-4 (a) Variation of Daily Mean Wind Speed in Şile, 1975-2010

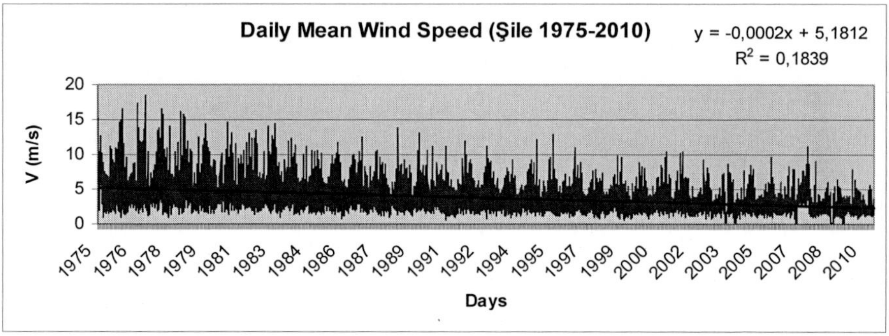


Figure 2-4 (a) gives the variation in daily mean wind speed in Şile. Temporal variations of wind speed values show a decreasing trend, with the significant level $\alpha$=0.01. Some statistical descriptions and histograms of mean monthly wind speed in Şile are given in Figure 2-4 (b).

1D Wavelet analyses of monthly mean wind speed in Şile is given in Figure 2-4 (c). A decreasing trend is observed at amplitude values. d8 explains the higher level fluctuations of long wave frequency, and d1 corresponds to lower layer signals (high frequency, short wave) fluctuations) of wind speed. d8 shows a periodic structure and a decreasing trend in recent years. Amplitudes at a high frequency signal have increased in recent years.

Figure 2-4 (d) presents analyses of continuous wavelet and localization of extreme values. In the first and last parts of the period, the role of large scale events can be observed. At the beginning, the frequency of large scale events varies between 5-12 months (semi-annual variations), while in recent years it varies between 5-40 months (inter annual variations). The same influences are also valid for energy potential variations at different levels in the study area.

Figure 2-4 (b) Statistical Properties, Relative and Cumulative Frequency Distributions, Monthly Mean Wind Speed in Şile (1975-2010)

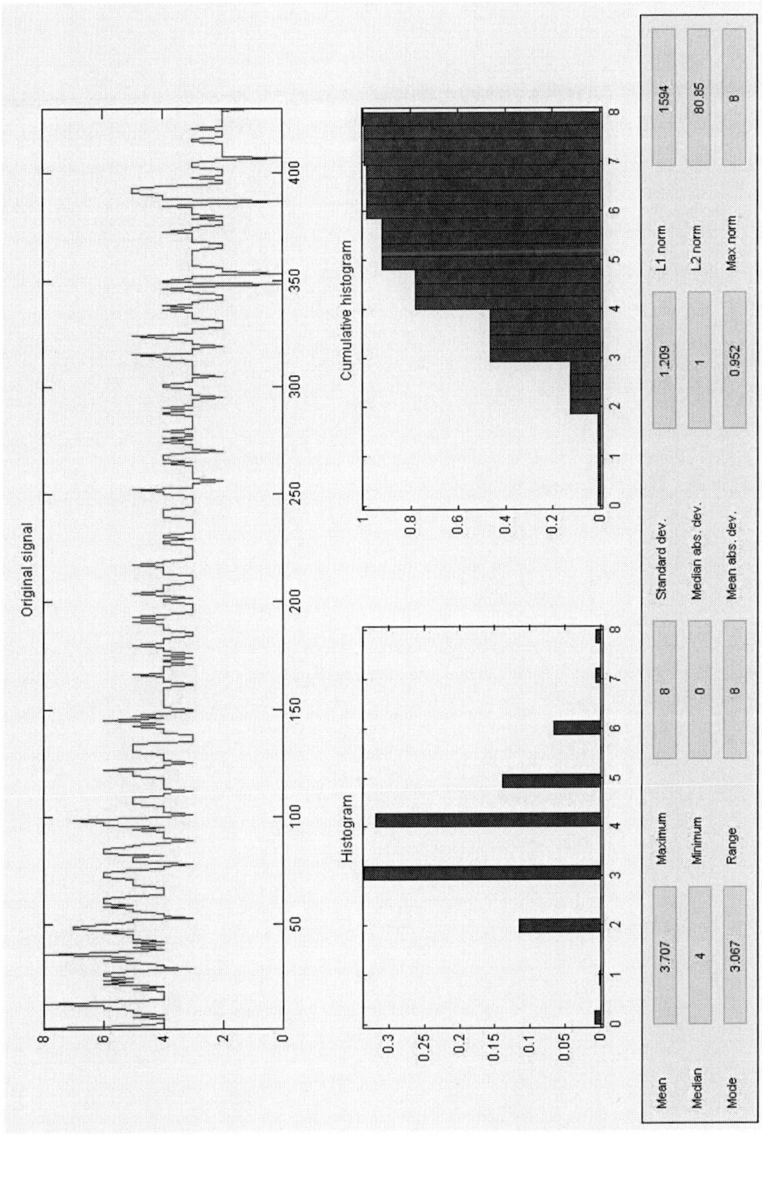

Figure 2-4 (c) 1D Wavelet, DMeyer, 8 Levels, (Monthly Wind Speeds, Şile, 1975-2010)

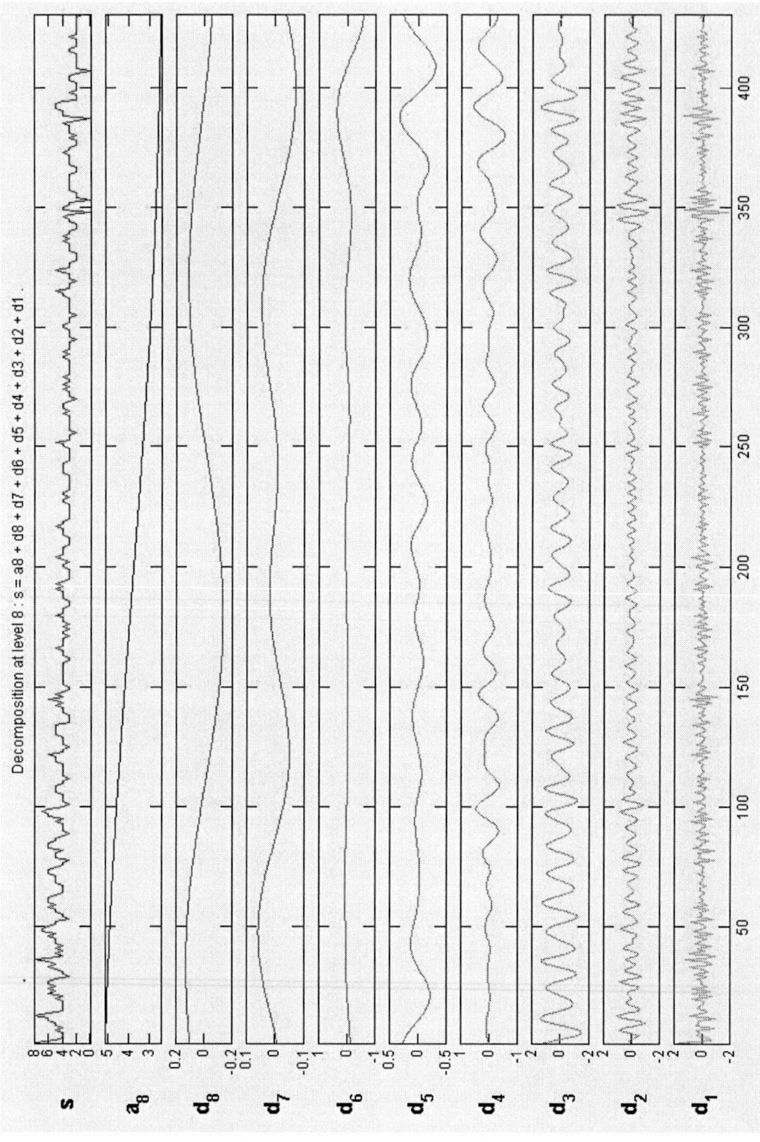

Figure 2-4 (d) Continuous Wavelet, DMeyer, Sampling: 1 (Monthly Wind Speeds, Şile, 1975-2010)

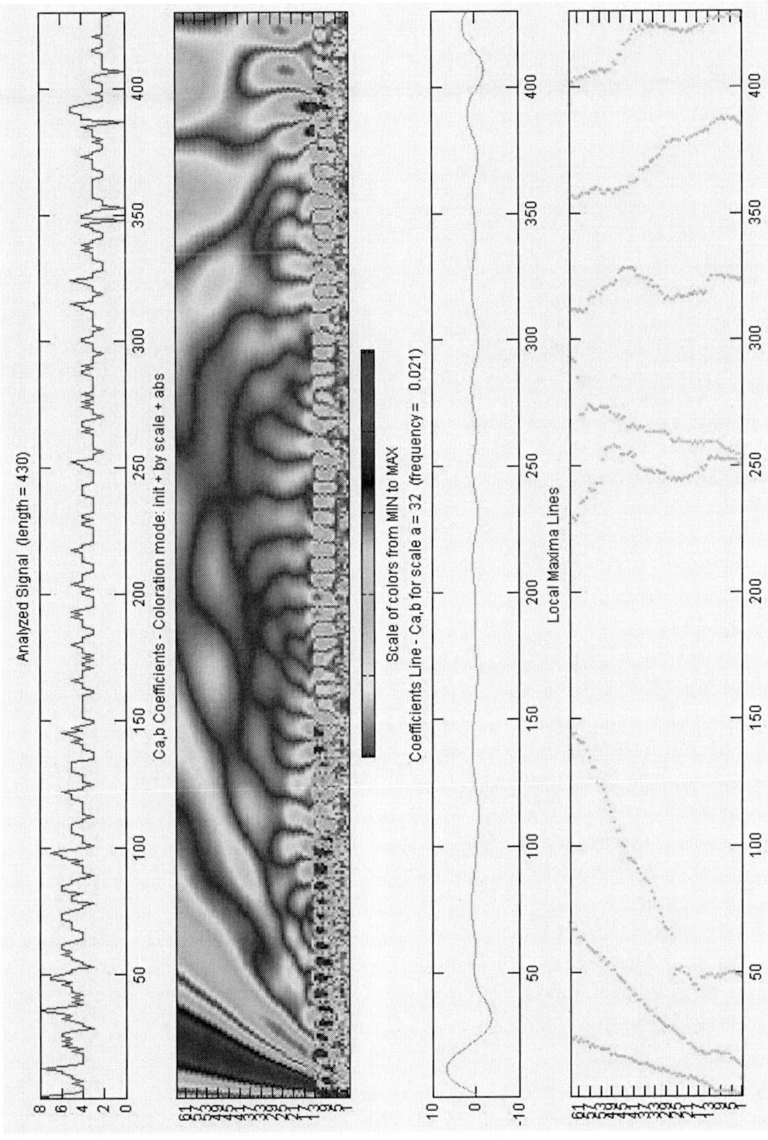

## 4. Results and Conclusion

Statistical and wavelet analyses of wind speed variations in Şile (Istanbul) between 1975 and 2010 explain wind speed variations, the estimation of wind energy potential, and the role of small and large scale events on wind speed and energy potential. There is significant evidence of decreasing wind speed in the study area. Faster urbanization is one of the reasons behind this trend, together with large scale periodicity.

This study is related to some national energy plans on wind energy converting systems (WECSs). Expected energy output was performed for different sizes. For instance, it is recommended that one large wind turbine would facilitate the construction of a water pumping system in and near the vicinity of Şile. These results are similar to energy potential analyses conducted in some parts of the Aegean Sea Regions, and will contribute to the decision-making and planning processes concerning national renewable energy.

## References

Akdağ, S. A.Dinler ve S. Menteş (2010), "Rüzar Karakteristiğinin Analizi", s. 23-29, İTÜ, http://www.emo.org.tr/ekler/3dbe3b58a1bafbb_ek.doc *(April 17, 2011)*

Akdağ, S., A, ve Ö. Güler (2008), "Weibull dağılım Parametrelerini Belirleme Metotlarının Karşılaştırılması", VII. Ulusal Temiz Enerji Sempozyumu, UTES'2008 17-19 Aralık 2008, İstanbul, s. 707.

Almalı, M. N., ve T. Ete (2009), "Ortalama Rüzgar Hızı ve Güç Yoğunluğunun Hesaplanması", YEKSEK2009, 5. Yenilenebilir Enerji Kaynakları Sempozyumu, EMO Yayın No: SK/2009/7, ISBN978-9944-89741-9, s. 97-102.

Altunc, M., Tokgozlu, A., Goymen, H., Tolun, S., Aslan, Z., Unsalan, D. (2002), "Comparison of wind energy potential in TRNC", International Conference on the Environmental Problems of the Mediterranean Region, Near East University, Nicosia, TRNC.

Aslan, Z., ve H. Gençoğlu (2005), "Rüzgar-Güneş Enerji Sistemlerinden Yararlanma Olanakları", V. Ulusal Üretim Arastırmaları Sempozyumu, Istanbul Ticaret Üniversitesi, 25-27 Kasım 2005, s. 547

Aydın, N. Y. (2009), "GIS Based Site Selection Approach for Wind and Solar Energy Systems: A Case Study from Western Turkey", MS Thesis, Middle East Technical University, Graduate School of Natural and Applied Sciences, Ankara.

Can, Z., Aslan, Z., Oguz, O. and Siddiqi, A., 2005, "Wavelet transforms of meteorological parameters and gravity waves", Annals Geophysics, Vol. 23, pp. 650-663.

Kenisarina, M., V. M. Karslı and M. Caglar (2006), "Wind power engineering in the world and perspectives of its development in Turkey, Renewable and Sustainable Energy Reviews, 10, pp. 341-369

Meyer, Y. (2000), "The role of oscillation in some nonlinear problems", School on Mathematical Problems in Image Processing, ICTP, SMR 1237/4, pp. 4-22.

Nemes, C. and F. Munteanu (2011), "The Wind Energy System Performance Overview Capacity Factor vs. Technical Efficiency", International Journal of Mathematical Models and Methods in Applied Sciences, Issue 1 Vol. 5.

Özdamar A., Kavas M.G. (1999), "Rüzgar türbini pervanesi üzerine bir araştırma", Güneş Günü Sempozyumu, Bildiriler Kitabı, pp. 151-160, 1999.

Shada, A. (2008), "Theoretical investigation and mathematical modeling of a wind energy system Case study for Mediterranean and Red Sea", MS. Thesis, Berlin Technical University, Faculty of Technique and Informatics, Berlin.

Siddiqi, A., Aslan, Z. and Tokgozlu, A. (2002), "Wavelet based computer simulation of some meteorological parameters: Case study in Turkey", Trends in Industrial and Applied Mathematics. H. Siddiqi and M. Kocvara (Editors), pp. 95-105, Kluwer Academic Publishers, London.

Siddiqi, A., Alsan, S., Rasulov, M., Oğuz, O. and Aslan Z. (2005), "International workshop on applications of wavelets to real world problems: IWW2005", July 17-18, ISBN: 975-6516-11-9, ITICU Publication No: 12, Istanbul, p. 290.

Tokgozlu, A., O. Kaynar, M. Altunç, Z. Aslan (2010), "Wind Speed and Energy Potential Analyses", Wavelet Symposium 2010, IAU, Istanbul

Tolun S., Mentes S., Aslan Z., Yukselen M., 1995, "The Wind Energy Potential of Gokceada in the Northern Aegean Sea", 1995, Renewable Energy, Vol. 6, No. 7, pp. 679-685

Triola, M. F. (1998), Elementary Statistics, Seventh Ed. Addison-Wesley Longman Inc., 791, NY.

Turksoy F. (1995), "Investigation of wind power potential at Bozcaada, Turkey", Renew Energy, 6(8), pp. 917-23.

# Chapter Three:

# Energy:
# Impacts of Global Climate Change

# Sustainable Development and Turkey's Biomass Energy Potential*

Coşkun Karaca[1]
and M. Mustafa Erdoğdu[2]

## 1. The Energy Situation in Turkey: An Introduction

Fossil fuels are the energy source of our time.[3] According to UNEP (2012: 1), 80 per cent of current global energy needs, or 66 per cent of power supply, are fossil fuel-based. However, the energy from fossil fuels has a major negative impact on both the environment and on human health. Global energy systems currently represent some 60 per cent of total current greenhouse gas (GHG) emissions. Hence, an energy policy for a sustainable future needs to be based on high levels of energy efficiency and greater use of renewable energy.

Turkey is the sixth largest economy in Europe and has been growing quite rapidly in recent years. Historic trends suggest that with sustained economic growth, energy consumption (and electricity consumption in particular) will increase at a similar pace to GDP. In the coming decade, it is estimated that primary energy consumption will increase by around 4 per cent per annum. By 2020, it is estimated that electricity consumption

---

* This is a revised version of a paper which was presented at the First International Symposium on Climate Change, Environment and Energy, organised by Istanbul Aydin University, on the theme of Global Climate Change, Environment and Energy: Global Challenges & Opportunities to Global Stability on 25 April 2011 in Istanbul, Turkey.

[1] Assistant Professor, Department of Public Finance, Faculty of Economic and Administrative Sciences, Cumhuriyet University, 58100/Sivas, TURKEY, Tel: 0 (322) 338 72 63, ckaraca@cumhuriyet.edu.tr

[2] Professor, Department of Public Finance, Faculty of Economics, Marmara University, Göztepe Kampusü 34722/Istanbul, TURKEY, Tel: 0 (216) 414 05 45, mustafaerdogdu@marmara.edu.tr

[3] Fossil fuels are energy resources like coal, oil, and natural gas. They are organic materials that form over millions of years after dead plants and animals are buried and exposed to extreme pressures and temperatures.

will have increased in the range of 6.7 per cent (low) to 7.5 per cent (high). The capital investment required to meet this growth in energy consumption over the period 2010-2030 is estimated at between US$225-280 billion (WEC, 2012: 13). The characteristics of the Turkish Energy Sector can be summarised as follows:

- A high increase in absolute energy consumption.
- High dependence on imported energy.
- Energy intensive industrial consumption.

Turkey is a very vulnerable country in terms of energy. This is because the overwhelming share of Turkey's energy requirement is met by fossil fuels, and 90 per cent of such energy is imported. As a result, Turkey is exposed to energy security challenges, such as fuel supply disruption and fuel price volatility. This situation provides a backdrop for the development of renewable energy, which offers the benefit of reducing greenhouse gas emissions while also reducing the soaring dependence on imported energy, predominantly in the form of oil and natural gas. Turkey's energy policy requires radical reform, not only because of its energy dependency, but also because of worsening environmental pollution. Turkey's greenhouse gas emissions were about 114 per cent higher in 2010 than in 1990. Thus, renewable energy resources are becoming more important, due to increasingly restrictive environmental constraints.

As a source of renewable energy, biomass is a very versatile material and is therefore the chosen focus of this paper. Unlike other natural resources, such as petroleum, coal, and nuclear fuels, biofuels are a renewable energy source based on the carbon cycle. Therefore, biomass fuels can contribute to waste management, improve fuel security, and help to prevent or slow down climate change, although they are not by themselves a comprehensive solution to these problems. From this starting point, the current study focuses on the level of Turkey's biomass energy potential, and how to make sufficient use of this potential.

In the light of the above points, this paper seeks to contribute to the understanding of how various climate policy interactions can affect the use of biomass for energy. In particular, the present work seeks to assess how the combination of national and international climate policy instruments impacts upon biomass use. The paper is structured as follows: the next section will look at the issue of energy from a sustainable development perspective; parts 3 and 4 will identify the advantages and disadvantages of biomass energy; part 5 will investigate biofuel energy potentials of

Turkey; part 6 will contain a discussion of different policy instruments for biofuel energy promotion and provide policy suggestions; the last part concludes.

## 2. Biomass Energy from the Perspective of Sustainable Development

The scientific consensus is that human-induced GHG emissions are the primary cause of global warming, and that carbon dioxide is the most important of the anthropogenic GHGs. Human activities are adding $CO_2$ to the atmosphere far faster than natural processes can remove it. Emissions have been, and continue to be, driven by economic growth. Growth in the production of goods is damaging many aspects of the physical environment. The atmosphere, soil, water supplies, and plant and animal life are all being adversely affected. Air, water, and marine pollution are reducing the quality of life of tens of millions of people, and desertification is claiming more and more fertile land (UN, 2007: 47). Sustainable development requires a sustainable supply of clean and affordable renewable energy sources that do not cause negative societal impacts. Renewable energy creates multiple public benefits, such as environmental improvement (reduction of power plant greenhouse emissions, thermal and noise pollution), increased fuel diversity, reduction of energy price volatility effects on the economy, national economic security (fossil energy is vulnerable to political instabilities, trade disputes, embargoes and other disruptions), and an increase in economic productivity and GDP through more efficient production processes. As a recent World Energy Council publication (2010a) points out, the supply of sustainable energy is one of the most significant agendas facing mankind over the coming decades. The situation clearly demands a greater focus on renewable energy resources and green growth policies. Such policies can stimulate economic growth while preventing environmental degradation, biodiversity loss and the depletion of unsustainable natural resources, conserving them for future generations. Many countries, including Turkey, have made pledges under the Copenhagen Accord to reduce greenhouse-gas emissions. To meet its target, Turkey needs to increase the share of renewable energy resources in the total energy supply. Turkey's great potential for hydro, biomass, wind, geothermal and solar energy sources is an asset. Among the renewable energy sources available, biomass is particularly important, because of its immediate potential.

Biomass is derived from recently living organisms, which include plants, animals, and their by-products. Manure, garden waste, and crop residues are all sources of biomass. The oil crises of the 1970s prompted interest in biofuels[4] in many countries, as an alternative to fossil fuels for use in transportation. Brazil accelerated its national ethanol programme (Proálcool) after oil prices peaked in 1979, the United States (US) launched a corn-based ethanol programme at almost the same time, but on a smaller scale than Brazil, and Germany has been a leader in the large-scale production of biodiesel fuel from rapeseed and sunflower seed, crops commonly used to produce vegetable oil for human consumption (Worldwatch, 2006: 5).

Biofuels are combustible materials directly or indirectly derived from biomass, commonly produced from plants, animals and micro-organisms, but also available from organic waste. Biofuels for transport are commonly categorised, according to their current or future availability, as first, second, or third generation biofuels. Second and third generation biofuels are also called "advanced" biofuels. This differentiation is not always straightforward, due to overlaps regarding feed-stocks and processing technologies, as well as uncertainties regarding environmental impacts (UNEP, 2009: 25). First-generation biofuels are commercially produced using conventional technology. The basic feed-stocks are seeds, grains, or whole plants from crops such as corn, sugar cane, rape seed, wheat, sunflower seeds, or oil palm. The most common first-generation biofuels are bioethanol (currently over 80 per cent of liquid biofuels production by energy content), followed by biodiesel, vegetable oil, and biogas. Second-generation biofuels can be produced from a variety of non-food sources. These include waste biomass, the stalks of wheat, corn stover (stalks, leaves, husks, and cobs), wood, and special energy or biomass crops. Algae fuel, also called oilgae, is a biofuel from algae that falls into the third-generation biofuel category. Algae are feed-stocks from aquatic cultivation for production of triglycerides (from algal oil) used to produce biodiesel. Other third generation biofuels include alcohols like biopropanol or biobutanol, which, due to lack of production experience, are usually not considered likely to become relevant fuels on the market before 2050 (IEA, 2008).

[4] Only liquid transport fuels are considered here, although there are other fuels derived from biomass.

**Table 3-1 The Categories of Land-based Biomass Supply for Energy Production**

| Biomass category | Comment |
|---|---|
| **Category 1.** Residues from agriculture | By-products associated with food/fodder production and processing, both primary (e.g., cereal straw from harvesting) and secondary (e.g., rice husks from rice milling) residues. |
| **Category 2.** Dedicated biomass production on surplus agricultural land | Includes both conventional agriculture crops and dedicated bioenergy plants including oil crops, lignocellulosic grasses, short-rotation coppice and tree plantations. |
| **Category 3.** Dedicated biomass production on marginal lands | Refers to biomass production on deforested or otherwise degraded or marginal land that is judged unsuitable for conventional agriculture but suitable for some bioenergy schemes (e.g., via reforestation). |
| **Category 4.** Forest biomass | Forest sector by-products including both primary residues from silvicultural thinning and logging, and secondary residues such as sawdust and bark from wood processing. |
| **Category 5.** Dung | Animal manure. |
| **Category 6.** Organic wastes | Biomass associated with materials use, for example, organic waste from households and restaurants and discarded wood products including paper, construction and demolition wood; availability depends on competing uses and implementation of collection systems. |

Source: IPCC, 2012, s. 224.

Figure 3-1 Shares of Global Primary Biomass Sources for Energy

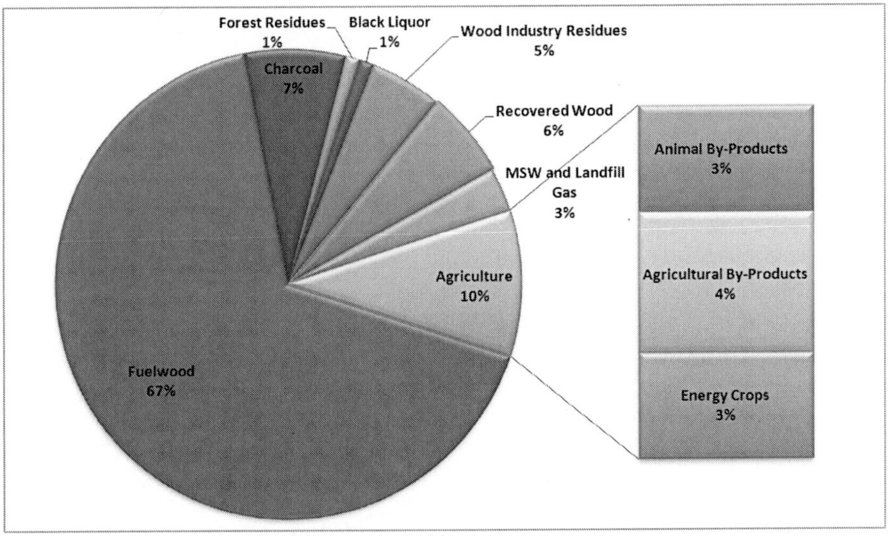


Source: IPCC, 2007a, b; IEA Bioenergy, 2009

Biomass in the form of fuel wood and animal wastes is the main fuel for heating and cooking in many urban and rural areas in Turkey (Koyun, 2007: 6), and has a substantial potential to contribute to the country's increasing energy demands. Bioenergy appears to have formidably positive environmental properties, as there is no net release of carbon dioxide and very-low sulphur content resulting from its use. The next section will document some advantages and disadvantages of biomass energy.

## 3. Advantages and Disadvantages of Biomass Energy

Though viewed by many as a "new" energy resource, wood and other forms of biomass provided the majority of energy and fuel needs until about 150 years ago. Subsequently, fossil fuels became the primary energy source in industrialised countries and biomass was gradually sidelined. However, according to UNEP, over 500 million households in developing countries still use traditional biomass for cooking and heating. Skyrocketing and volatile oil prices since the 1970s, coupled with increasing concerns related to the environment since the 1980s, have motivated the search for alternative sources of energy that are economically efficient, socially equitable, and environmentally sound. Consequently, biofuels, produced from biomass feed stocks, have come to the fore and have received considerable attention from decision-makers in government and industry, as well as the wider global public. Biofuels are characterised (by those who believe in their considerable potential) as an automotive fuel with a bright future, and as representing a central technology in the fight against climate change. Others see biofuels as a diversion from the tough climate mitigation actions needed, or as a threat to food security and thus a key challenge in the achievement of the poverty-related Millennium. The truth, which clearly depends on the individual circumstance of different countries and regions, must lie somewhere in between.

Biofuel production costs are currently higher than those of fossil fuels, if externalities are left out of account. According to Kaygusuz and Keleş (2012: 90), biofuels are currently about 2.3 times more expensive than fossil fuels, and this high cost is critically related to the cost of raw materials, not to mention the high costs of collection, transport, and storage. The large volumes of water required to produce biomass also need to be considered. As Demirbaş (2011: 17) points out, the cost varies depending on the feedstock, geographic area, methanol prices, and seasonal variability in crop production.

Major risks related to bioenergy development have been identified: these include the impact of raw materials price increases due to competition from a growing bioenergy industry, as well as increasing land use competition linked to bioenergy production (Bowyer, 2011: 30). The United Nations Food and Agriculture Organisation estimated in 2008 that biofuels accounted for approximately 10 per cent of the recent food price increases around the world (WEC, 2010a). Food versus fuel is a real dilemma, with regard to the risk of diverting farm land or crops for liquid biofuels production to the detriment of the food supply on a global scale (Demirbaş, 2011: 17). Another identified risk is the expansion of biofuel feedstock production into areas that provide valuable ecosystems that support high biodiversity and services, crucial to our economies and to human life (UNEP-WCMC, 2009). Moreover, the abundant use of fertilisers and manure for bioenergy production may also result in considerable environmental problems.

After a substantial amount of research and experimentation, scientists and researchers have come close to concluding that biomass energy has many advantages over fossil fuels, and that it helps in reducing the greenhouse effect. However, as a recent United Nations Environment Programme publication pointed out, it is important to recognise that not all biofuels perform equally well in terms of their impact on climate, energy security, and ecosystems. Environmental and social impacts need to be assessed throughout the entire life-cycle (UNEP, 2009: 15).

Biomass has recently become increasingly popular as an alternative energy source for a variety of reasons, including the following:

- It is a renewable source of energy and available in unlimited quantities.
- It helps in reducing the greenhouse effect, provides scope for using wastes and residues, reducing waste disposal problems, and consequently enhances the environment and public health.
- It reduces energy dependence and improves trade balances by substituting imported domestic biomass for fossil fuels.
- It creates new jobs and provides opportunities for economic and social development in rural communities.

Now, let us look at these advantages more closely. First of all, biomass is a renewable source of energy with a relatively small carbon footprint when compared to fossil fuel. Biomass absorbs carbon dioxide during growth and emits it during combustion; therefore, it recycles atmospheric carbon and does not add to the greenhouse effect. Because it is available in unlimited quantities for an unlimited period of time, there should be no shortage in the availability of biomass as a fuel. Key to the introduction of biomass production in large amounts is the rationalisation of agriculture, especially in developing countries. There is room for considerably higher land use efficiencies that can more than compensate for the growing demand for food.

The other important advantage of biomass energy over fossil fuels is that it helps in reducing the greenhouse effect. The burning of fossil fuels adds new $CO_2$ to the atmosphere, while the burning of biomass balances the $CO_2$ absorbed by plants during growth. Fossil fuels do not only produce $CO_2$, but also sulphur dioxide and lead oxide during their burning, both of which are very toxic gases. The use of biofuels, however, prevents the release of sulphur dioxide into the atmosphere. Biomass also helps in controlling methane, another greenhouse gas. In this context, the findings of the International Energy Agency are shown in Figure 3-2 below.

Figure 3-2 Lifecycle GHG Emissions per unit of Output for a Range of Bioenergy (green) and Fossil (black) Options

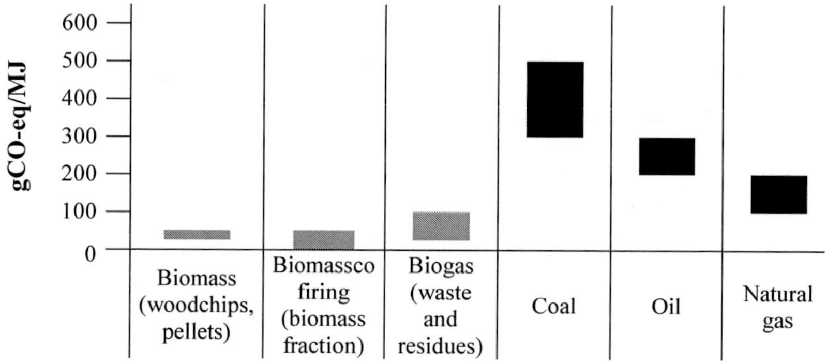


Source: IEA, 2012, p. 17.

Bioenergy systems offer significant possibilities for reducing greenhouse gas emissions by replacing fossil fuel with biomass for energy production. In some instances, it is claimed that carbon sequestration to plant and soil, combined with non-invasive farming methods, make biomass electricity carbon negative; overall, less carbon is emitted than is removed from the atmosphere (IEA, 2010a). Many authors assert carbon neutrality, with emissions from combustion balanced by carbon capture of the next crop.[5]

The net carbon emission of producing one unit of bioenergy is 10-20 folds less than carbon emissions from fossil fuel-based products. Using and producing biodiesel causes 41-78 per cent less emission than petroleum-based diesel (Özertan, 2008: 25). In support of this proposition, the findings of Wu and Ma (2003: 145) indicate that using 1.00 x$10^4$ tons of biomass plant to substitute for coal in electricity generation would reduce carbon dioxide ($CO_2$) by 1.40 x$10^4$ tons, sulphur dioxide by 40 tons, and the ash emitted into the atmosphere by 100 tons. According to Wihersaari (2005: 435-43), other greenhouse gases such as methane and nitrous oxide are emitted as 2 per cent or less of total emissions. A further clear advantage of biomass energy over fossil fuels is that its production helps in cleaning the surrounding area. Biomass can utilise a lot of municipal waste, in the form of garbage, paper, water, industrial scrap, etc. Thus, using biomass as a fuel not only gives us energy but also helps to remove garbage from the surroundings and to keep the area clean.

Another great advantage of biomass energy is that it is an indigenous fuel. The fuels from biomass materials can be produced locally and no high technology is required. Producing fuel from biomass materials reduces the dependence of a country on foreign resources for their fuel requirements. Moreover, since this indigenous fuel is labour intensive, it can contribute to the generation of new jobs,[6] particularly in rural and farming communities. The number of employed workers required is 11-15 times greater than the workers in nuclear energy production, and 3-6 times greater than the fossil energy production in the associated processes (Duygu, 2004: 466). Some other socio-economic benefits can be counted, such as slowing down migration from rural areas to cities, decreasing the

---

[5] For these contributions, see, P. Lettens, B. Muys, R. Ceulemans, E. Moons, J. Garcia and P. Coppin (2003). Energy budget and greenhouse gas balance evaluation of sustainable coppice systems for electricity production, Biomass Bioenergy, 24, 179-197; Vande Walle, I. Van Camp, N. Van de Casteele, L. Andrheyen K. and Lemeur, R. (2007). Short-rotation forestry of birch, maple, poplar and willow in Flanders (Belgium). II. Energy production and $CO_2$ emission reduction potential, Biomass Bioenergy, 31, pp. 276-283; Matthews, W. (2001). Modelling of energy and carbon budgets of wood fuel coppice systems, Biomass Bioenergy, 21, pp. 1-19.

[6] The Green Economy Report modelling indicates that if around half of total energy investments were directed towards energy efficiency and renewable energy (including the expansion of second generation biofuels), the result would be 20 per cent more employment than business as usual by 2050, while delivering robust economic growth and reduced emissions (UNEP, 2012: 1).

issues associated with rapid urbanisation, and developing a biodiesel production industry in the country (Adıyaman and Günay, 2008: 120-121). Therefore, it can be argued that biomass energy production would help to decrease unemployment and to revitalise rural communities. In this context, we can say that biomass acquisition, transportation, and use in power generation serves to promote greater stability in society. According to Table 3-2, in the year 2020, 70 per cent of Turkey's energy demands will be met through imports. With high economic growth expected to resume after the 2008/09 global economic crisis, and without sustained policy shifts in the meantime, Turkey faces the prospect of risks related to electricity security and limited access to competitive oil and gas supplies, which could significantly reduce its growth potential (IEA, 2009: 20).

Considering the volume of energy demand, ensuring sufficient energy supply has been and remains the Turkish government's main energy policy concern. Since the government wants to minimise fluctuations in energy generation, it plans to generate between 5-10 per cent of its energy needs from nuclear technology over the next decade. However, as the recent Fukushima nuclear disaster in Japan has demonstrated, this could well be a very costly adventure for both the people who live in Turkey and their international neighbours, because Turkey is in a high-risk earthquake zone.

Biomass, on the other hand, represents a secure domestic source of energy that is not subject to the price fluctuations and supply uncertainties of imported petroleum and natural gas. Unlike other non-renewable forms of energy, biomass energy can be produced and consumed in a sustainable fashion, and results in no net contribution of carbon dioxide to global warming. Technologies have advanced so as to be largely pollution free. There is zero to little production of slag, ash, $SO_X$, $NO_X$ or $CO_2$ in biomass consumption. However, this energy form is not completely problem-free either. Although biomass energy sources such as biofuel are cleaner than fossil fuels, they carry some environmental risks. Moreover, the use of appropriate modern technology is critical and costly (Balat, 2005: 32-41; Demirbaş, 2001: 1238-59).

**Table 3-2 Energy Balances and Net Energy Imports in Turkey (Mtoe)**

| | | 1973 | 1990 | 2000 | 2005 | 2007 | 2008 | 2009 | 2010 | 2020 |
|---|---|---|---|---|---|---|---|---|---|---|
| **TOTAL PRODUCTION** | | **15.5** | **25.8** | **25.9** | **23.9** | **27.3** | **29.0** | **30.3** | **32.5** | **66.0** |
| **Coal** | | 5.2 | 12.4 | 12.5 | 10.8 | 14.8 | 16.7 | 17.4 | 17.5 | 66.0 |
| **Oil** | | 3.6 | 3.6 | 2.7 | 2.2 | 2.1 | 2.1 | 2.3 | 2.7 | 0.7 |
| **Gas** | | - | 0.2 | 0.5 | 0.7 | 0.7 | 0.8 | 0.6 | 0.6 | 0.2 |
| **Biomass** | | 6.5 | 7.2 | 6.5 | 5.4 | 5.1 | 4.9 | 4.7 | 4.6 | 3.9 |
| **Nuclear** | | - | - | - | - | - | - | - | - | 8.2 |
| **Hydro** | | 0.2 | 2.0 | 2.7 | 3.4 | 3.1 | 2.9 | 3.1 | 4.5 | 9.4 |
| **Wind** | | - | - | - | 0.0 | 0.0 | 0.1 | 0.1 | 0.3 | 0.7 |
| **TOTAL NET IMPORTS** | | **8.7** | **27.8** | **50.0** | **59.7** | **73.8** | **70.6** | **71.4** | **70.2** | **151.8** |
| **TOTAL NET IMPORTS (%)** | | **35.8%** | **52.6%** | **65.5%** | **70.7%** | **73.8%** | **71.6%** | **69.18%** | **72.52%** | **69.7%** |
| **Coal** | Exports | - | - | - | - | - | - | - | - | - |
| | Imports | 0.0 | 4.2 | 9.3 | 11.7 | 14.6 | 12.9 | 13.3 | 13.7 | 43.5 |
| **Oil** | Exports | 0.8 | 1.9 | 1.3 | 5.4 | 6.1 | 6.5 | 6.0 | 7.3 | - |
| | Imports | 9.7 | 23.1 | 30.5 | 33.5 | 37.6 | 36.0 | 33.9 | 38.6 | 60.2 |
| **Gas** | Exports | - | - | - | - | 0.0 | 0.4 | 0.6 | 0.6 | 0.7 |
| | Imports | - | 2.7 | 12.1 | 22.1 | 29.8 | 30.6 | 32.8 | 34.8 | 52.0 |

Source: IEA (2009); ETKB (2012).
Note: The figures for 2020 are forecast.

The costs of biomass fuel vary considerably. The crop types, agricultural practices, land and labour costs, plant size, processing technologies and government policies in different regions mean that ethanol production costs and prices vary considerably from region to region. According to Demirbaş (2011: 17), the cost of producing bioethanol in a dry mill plant currently totals US$1.65/gallon, and the largest ethanol cost component is the plant feedstock. It has been shown that plant size has a major effect on cost. The plant size can reduce operating costs by 15-20%, saving another $0.02-$0.03 per litre. Thus, a large plant with production costs of $0.29 per litre may save $0.05-$0.06 per litre over a smaller plant. P Lange (2007: 45) indicates that the production of biofuel using today's technology costs 60-120 dollars per barrel. When we take into account the high price of crude oil (86 dollars per barrel at the time of writing, in October 2012) and its external cost on human health and the environment, it would seem an acceptable option to choose biofuel over petroleum.

The costs on the table represent the upper and lower bounds of direct costs that occur in the case of obtaining 1 kWh of energy from renewable and fossil energy resources. These costs do not include negative externalities.[7] Indeed, the external costs of fossil fuels are very high when compared to biomass. As reported by Research Reports International (2006), "The lack of policy to credit the distinct sustainability benefits of biomass or to require sustainable use of natural gas and other fossil resources makes the cost of biomass energy to appear high. In order for biomass-fuelled generation to be able to compete effectively on a large scale with other renewable and fossil fuel alternatives, its environmental benefits must be realised".

When comparing fossil fuels and biomass, two aspects are very important; the first is that when produced via sustainable means, biomass emits roughly the same amount of carbon during conversion as is taken up during plant growth; the second is that using harmful wastes for energy production provides a dual benefit. In other words, almost no production costs exist for biomass energy, but there are additional gains when external costs are taken into account. The difficulty of calculating the external costs of greenhouse gases generates the delusion that fossil fuels are cheaper than biomass energy.

As Table 3-4 shows, the direct cost of electricity production is greater when it is produced through biomass than with fossil fuels. However,

[7] "Externalities occur where market-priced transactions do not fully incorporate all the benefits and costs associated with transactions between economic agents" (Stewart and Gani, 1991, p. 569).

since Turkey is an agricultural and stockbreeding country, in the long run it has a good potential to replace crude oil and natural gas with biomass for the purpose of electricity production (Dündar and Arıkan, 2004). It is worth noting that the 2011 cost of Turkey's imported energy was over US$54 billion. This figure means that close to 40 per cent of Turkey's annual exports are utilised to fund energy imports. This will mean that future policies will trend towards the development of generation technologies using indigenous energy resources that can achieve the aims of supply security and sustainability (WEC, 2012: 13).

**Table 3-3 Levelled Cost of Electricity (LCOE) from Different Energy Sources**

| | | | EC (2008) | EPRI (2008) | House of the Lords (2008) | MIT (2009) | Min. ($/kWh) | Max. ($/kWh) |
|---|---|---|---|---|---|---|---|---|
| Nuclear | Overnight cost | $/kW | 2.552-4.378 | 3.980 | 3.000 | 4.000 | | |
| | Capacity factor | | 85% | 90% | 77% | 85% | | |
| | **LCOE** | **$/MWh** | **65-110** | **73** | **90** | **84** | **0,06** | **0,11** |
| Coal | Overnight cost | $/kW | 1.295-1865 | 2.45 | 2.14 | 2300 | | |
| | Capacity factor | | 85% | 80% | 81% | 85% | | |
| | **LCOE** | **$/MWh** | **52-65** | **64** | **82** | **62** | **0,05** | **0,08** |
| Gas | Overnight cost | $/kW | 622-946 | 800 | 1.046 | 850 | | |
| | Capacity factor | | 85% | 80% | 81% | 85% | | |
| | **LCOE** | **$/MWh** | **65-78** | **73-87** | **78** | **65** | **0,06** | **0,08** |
| Biomass | Overnight cost | $/kW | 2.617-6.580 | 3.235 | 3.674 | | | |
| | Capacity factor | | 85% | 80% | 80% | | | |
| | **LCOE** | **$/MWh** | **104-253** | **73-86** | **180** | | **0,07** | **0,25** |
| Onshore wind | Overnight cost | $/kW | 1.295-1.775 | 1.995 | 2.222 | | | |
| | Capacity factor | | 23% | 33% | 27% | | | |
| | **LCOE** | **$/MWh** | **97-142** | **91** | **146** | | **0,09** | **0,14** |
| Offshore wind | Overnight cost | $/kW | 2.267-3.562 | 1.995 | 3.148 | | | |
| | Capacity factor | | 39% | 33% | 37% | | | |
| | **LCOE** | **$/MWh** | **110-181** | **91** | **162** | | **0,09** | **0,18** |
| Hydro | Overnight cost | $/kW | 1.166-8.549 | | | | | |
| | Capacity factor | | 50%-57% | | | | | |
| | **LCOE** | **$/MWh** | **45-240** | | | | **0,04** | **0,24** |
| Solar PV | Overnight cost | $/kW | 5.311-8.938 | | | | | |
| | Capacity factor | | 11% | | | | | |
| | **LCOE** | **$/MWh** | **674-1.140** | | | | **0,67** | **1,14** |
| Solar thermal | Overnight cost | $/kW | 5.181-7.772 | 4.6 | | | | |
| | Capacity factor | | 41% | 34% | | | | |
| | **LCOE** | **$/MWh** | **220-324** | **175** | | | **0,22** | **32,4** |

**Notes:** 1. All values are in USD 2007. The data reported in the studies in different currency and/or different years were converted into USD 2007 using the annual average exchange rate and a 2 per cent annual inflation rate. 2. All data are exclusive of carbon penalties. For more information, see IEA, Projected Costs of Generating Electricity, International Energy Agency, 2010b, s. 191-192.

**Table 3-4 Current and Projected Estimated Production Costs and Efficiencies of Bioenergy Chains at Various Scales in World Regions for Power, Heat, and Biomethane from Waste (taken directly from available literature data)**

| Feedstock/ Country/ Region | Major Process | Efficiency, Application and Production Costs; Eff. = bioenergy/biomass energy Component costs in USD2005/GJ | Estimated Production Costs US $cents_{2005}$/kWh | Potential Advances US $cents_{2005}$/kWh |
|---|---|---|---|---|
| Wood log, residues, chips/ Ag. Wastes/ Worldwide | Co-combustion with coal | 5 to 100 MWe, Eff. ~30 to 40%.[1,2] >50 power plants operated or carried on experimental operation using wood logs/residues, of which 16 are operational and using coal. More than 20 pulverized coal plants in operation.[3] Wood chips (straw) used in at least 5 (10) operating power plants in co-fi ring with coal.[3] | **2.9 – 5.3 $cents_{2005}$/kWh** Inv. Cost (USD/kW): 100 – 1,300[1] | Reduce fuel cost by improved pretreatment, characterization and measurement methods.[4] Torrefied biomass is a solid uniform product with low moisture and high energy content and more suitable for co-firing in pulverized coal plants.[3] Cost reduction and corrosion-resistant materials for coal plant needed.[5] |
| Wood log, residues, chips/ Ag. Wastes/ Worldwide | Direct combustion | 10 to 100 $MW_e$, Eff. ~20 to 40%.[1,2] Well deployed in Scandinavia and North America; various advanced concepts give high efficiency, low costs and high flexibility.[2] Major variable is biomass supply costs.[2] | **7.2 – 9.2 $cents_{2005}$/kWh** Inv. Cost (USD/kW): 1,600 – 2,500[1] | U.S. 2020 cost projections:[6] **6.3 – 7.8 $cents_{2005}$/kWh** Stoker fired boilers **7.5 – 8.1 $cents_{2005}$/kWh** |
| Municipal solid waste/Worldwide | Direct combustion (gasification and co-combustion with coal | 50 to 400 $MW_e$, Eff. ~22%, due to low-temperature steam to avoid corrosion.[7,8] Commercially deployed incineration has higher capital costs and lower (average) efficiency.[2] Four coal-based plants co-fi re MSW.[3] | **3.3 – 9.4 $cents_{2005}$/kWh**[7] | New CHP plant designs using MSW are expected to reach 28 to 30% electrical efficiency, and above 85 to 90% overall efficiency in CHP.[8] |
| Organic Waste/ Municipal solid waste /Manures/ Sweden/ EU in expansion | Anaerobic co-digestion, gas clean up, compression, and distribution | Widely applied for homogeneous wet organic waste streams and waste water.[2] To a lesser extent used for heterogeneous wet wastes such as organic domestic wastes.[2] | Elec.: **17 – 21 $cents_{2005}$/kWh**[1] | Improvements in biomass pretreatment, the biogas cleansing processes, the thermophilic process, and biological digestion (already at R&D stage).[1, 9] |
| Manures/Food residues | Farms/Food Industry | Biogas from farm animal residues and food processing residues at 0.15 to 0.29 $MW_e$.[10] | Elec./**GJ: 70 – 89** Inv. Cost (USD/kW): 12000 – 15000[10] | Improved designs and reduction in digestion times. Improvements in the understanding of anaerobic digestion, metagenomics of complex consortia of microorganisms.[11] |

**Abbreviations:** Inv. = Investment; Elec. = Electricity.

**References:** IPCC, 2012, s. 242-243. 1. Bauen et al. (2009); 2. IEA Bioenergy (2007); 3. Cremers (2009); 4. Econ Poyry (2008); 5. Egsgaard et al. (2009); 6. NRC (2009); 7. Koukouzas et al. (2008); 8. IEA (2008); 9. Petersson and Wellinger, 2009; 10. Kuuva and Ruska (2009); 11. Cirne et al. (2007).

## 4. Turkey's Biofuel Energy Potential

Among the different forms of renewable energy, biomass has a great potential to provide improved rural energy services, based on forest and agricultural residues (Başçetinçelik et al., 2005). Biomass has played and will apparently continue to play an important role in the supply of renewable energy in Turkey. The available agricultural and animal residues have been estimated to be roughly equal to 22-27 per cent of energy consumption in Turkey (Demirbaş, 2008: 834). Biomass power generation is second to hydropower among renewable energies. According to some estimates, such as Demirbaş (2008), the potential quantity of all biomass energy is about 32 Mtoe. However, the total recoverable bioenergy potential is estimated to be 17.2 Mtoe (Balat, 2005). On the other hand, in a recent Biosynergy (2008) report, this potential is given as only 8.6 Mtoe.

**Table 3-5 Biomass Residues in Turkey in terms of Energy Content**

| Biomass Residue Type | Total Energy Content (Mtoe)[a] | Share of Total Residues (%) |
|---|---|---|
| Crop residues | 22.8 | 62.8 |
| Fruit residues | 7.5 | 20.7 |
| Poultry residues | 6.0 | 16.5 |
| Total | 36.3 | 100 |

**Source:** Biosynergy, 2008, p. 39.
[a] Mtoe: million tons of oil equivalent

Erdoğdu (2008) points out that there are substantial areas of abandoned agricultural land that are not managed and are becoming overgrown in Turkey. Creating demand for biomass fuel is likely to bring these areas back into economic exploitation. The table below shows the planned biomass energy production in Turkey.

**Table 3-6 Planned Biomass Energy Production (ktoe) in Turkey**

| Years | 2003 | 2005 | 2010 | 2015 | 2020 | 2025 | 2030 |
|---|---|---|---|---|---|---|---|
| Classic biomass | 6613 | 6495 | 5754 | 4790 | 4000 | 3345 | 3310 |
| Modern biomass | 658 | 766 | 1660 | 2430 | 3520 | 4465 | 4895 |
| Total biomass | 7271 | 7261 | 7414 | 7320 | 7520 | 7810 | 8205 |

**Source:** Demirbaş (2006).

As the table shows, an important transformation in the biomass energy sector is expected to occur over the coming decades. According to this forecast, there will be a substantial increase in energy production from modern biomass. The table below shows the composition of the current and expected energy supply in Turkey.

**Table 3-7 Total Energy Supply in Turkey (share %)**

| | 1973 | 1990 | 2000 | 2005 | 2007 | 2008 | 2009 | 2010 | 2020 |
|---|---|---|---|---|---|---|---|---|---|
| TOTAL ENERGY SUPPLY (Mtoe) | 24.4 | 52.8 | 76.4 | 84.4 | 100.0 | 98.6 | 106.1 | 109.2 | 217.8 |
| Coal | 21.1% | 32% | 30% | 27% | 29.4% | 29.9% | 29.0% | 28.3% | 37% |
| Oil | 51.3% | 44.4% | 39.8% | 34.1% | 30.7% | 30% | 28.7% | 26.7% | 25.8% |
| Gas | - | 5.4% | 16.5% | 27% | 30.4% | 30.6% | 30.9% | 32.0% | 23.7% |
| Biomass | 26.5% | 13.7% | 8.5% | 6.3% | 5.1% | 4.9% | 4.4% | 4.2% | 1.8% |
| Nuclear | - | - | - | - | - | - | | | 3.8% |
| Hydro | 0.9% | 3.8% | 3.5% | 4% | 3.1% | 2.9% | 2.9% | 4.2% | 4.3% |
| Wind | - | - | - | - | - | 0.1% | 0.1% | 0.3% | 0.3% |

**Source:** IEA (2009); ETKB (2012).
**Note:** The figures for 2020 are forecast.

As seen in Table 3-7, although biomass had an energy supply share of 26.5 per cent in 1973, this share had fallen to as low as 4.2 per cent by 2010. This reflects the gradual but dramatic decline in traditional biomass energy supply, despite the fact that there has been a substantial increase in modern biomass energy supply. Oil has also lost ground due to the diversification of energy production. While it had about a 50 per cent share of energy supply in the early 1970s, this had fallen to less than 30 per cent by the late 2000s.

The Turkish biofuels sector did not receive much attention from the government or the Turkish private sector until the economic crisis of 2000, which led to increasing interest in the biofuels market. There was rapid growth in investment in the biofuels sector in Turkey from 2000 to 2005. In 2005 and 2006, many biodiesel production facilities were established all across the country. However, as indicated by an USDA Foreign Agricultural Service report, the sector then shrank in 2006 in response to unfavourable changes to the regulatory environment and related tax laws. Currently, development of the biofuels sector in Turkey is blocked by bureaucratic procedures and high tax rates. In 2009, only 7,000 MT biodiesel was produced, despite an annual capacity of 1.5 million MT. Bioethanol production was recorded as just 40,000 MT in 2009 (USDA, 2010: 2).

**Table 3-8 Production of Crop Residues in Turkey (2011)**

| Crops | Residue | Plantation Area (ha) | Production (tons) | Available Residue (tons) | Availability (%) | Energy value (MJ/kg) | Total Energy Value (GJ) |
|---|---|---|---|---|---|---|---|
| **Wheat** | Straw | 8,096,000 | 21,800,000 | 3,270,000 | 15 | 17.9 | 58,533 |
| **Barley** | Straw | 2,868,833 | 7,600,000 | 1,140,000 | 15 | 17.5 | 19,950 |
| **Rye** | Straw | 127,653 | 365,750 | 54,863 | 15 | 17.5 | 960 |
| **Oats** | Straw | 85,862 | 218,040 | 32,706 | 15 | 17.4 | 569 |
| **Maize** | Stalks | 589,000 | 4,200,000 | 2,520,000 | 60 | 18.5 | 46,620 |
| **Rice** | Husk | 99,400 | 900,000 | 540,000 | 60 | 16.7 | 9,018 |
| **Tobacco** | Stalks | 97,179* | 55,000* | 33,000 | 60 | 16.1 | 531 |
| **Cotton** | Stalks | 542,000 | 2,580,000 | 1,548,000 | 60 | 18.2 | 28,174 |
| **Sunflower** | Stalks | 655,700 | 1,335,000 | 801,000 | 60 | 14.2 | 11,374 |
| **Groundnuts** | Shell | 25,471 | 90,416 | 72,333 | 80 | 20.7 | 1,497 |
| **Soybeans** | Straw | 26,420 | 102,260 | 61,356 | 60 | 19.4 | 1,190 |

**Source:** TUIK, 2012.
(*) Data are provisional

Biomass production in Turkey based on forest and agricultural residue has great potential to provide improved rural energy services. Many types of energy crops are suitable to be grown in Turkey. Chief among these are agricultural crops. The total annual field crop production and residues in Turkey is given in Table 3-8. It was calculated that the total energy value of the field crop residues was around 228 PJ for the production period 2002-2003. The major crops in the ratio of the total energy value are maize (33.4 per cent), wheat (27.6 per cent), and cotton (18.1 per cent) (Öztürk and Başçetinçelik, 2006: 316).

According to USDA (2010: 4), there are currently four bioethanol production facilities established in Turkey. However, only one of these actively operates. This facility uses mostly corn and (very rarely) wheat as a raw material. The total capacity of the sector is currently 160,000 MT, and the total production in 2009 was 40,000 MT. Although some research and development activities are being carried out by universities and by the Scientific and Technological Research Council of Turkey (TUBITAK), advanced biofuels are not as yet commercialised (USDA, 2010: 5).

In summary, Turkey has the appropriate conditions to produce oil seed crops, due to its climate conditions and soil structure. However, domestic production is currently not sufficient to meet national consumption demand. Nevertheless, under current projections, and given the potential for additional investments that utilise domestic resources, Turkey has the opportunity to meet a large proportion of its increasing energy demand with a reduced reliance on imported oil and natural gas.

# 5. A Brief History of Turkish Renewables Policy and Current Legislation on Bioenergy

Market based policies for renewables started in Turkey in 1984, with third party financing, excise, and sales tax exemptions. The build-own-transfer (BOT) and the build-own-operate (BOO) schemes were put in place in the same year, and major power projects with the main objective of attracting private investors were financed. BOT projects were granted a treasury guarantee. Although BOT and BOO approaches attracted significant investment, they also created large public obligations with the government covering the market risk through take-or-pay contracts. The BOT and BOO financing schemes ended in 2000, and were replaced in 2001 by financial incentives within the framework of the Electricity Market Law (Law Number 4628) (Erdoğdu, 2008).

The Turkish government has voiced strong support for increased electricity production from renewable sources. The Ministry of Energy set the goal of using renewable energy power generation to meet 30% of Turkey's electricity demand by 2023. The ministry has not given a breakdown of how it expects this energy to be supplied, but the majority would have to come from hydropower and biomass, as those are the most abundant sources of renewable energy in Turkey (USDA, 2010: 3).

Existing Turkish law and regulation with relevance to the use of renewable energy sources is limited to two pieces of legislation: the Electricity Market Licensing Regulation, and the Law on Utilisation of Renewable Energy Resources for the Purpose of Generating Electrical Energy Law (No. 5346), which entered into force in 2005. There is no existing legislation for biomass alone (Erdoğdu, 2008).

The aim of the Law on Utilisation of Renewable Energy Resources for the Purpose of Generating Electrical Energy is to increase the use of renewable energy sources for generating electrical energy and, in the process of achieving this goal, to diversify energy resources, reduce greenhouse gas emissions, assess waste products, protect the environment, and develop the necessary manufacturing sector to realise these objectives.

The Turkish government did not specify any criteria at the initial stages of the establishment of the Turkish biofuels sector. The definitions of biodiesel and bioethanol were first included in Turkish regulations under the Petroleum Market Law in 2003. As the Law was not specific to biofuels and was instead based on petroleum, biodiesel associations and other stakeholders pressured the government to regulate biofuels through a separate Renewable Energy Law. In 2004, two regulations were adopted, "Legislation on Technical Criteria to be Applied to the Petroleum Market" and "Petroleum Market Licensing Legislation." These defined biodiesel as a liquid fuel, and laid out licensing criteria for the export, distribution, transport and sale of the product. In 2005, standards for biodiesel were set by the Turkish Standards Institute in accordance with European Union Standards. Accordingly, Biodiesel could be marketed as either "automobile biodiesel" or "other fuel biodiesel." Between 2003 and 2006, many biodiesel production facilities were established. Following the introduction of mandatory application for licensing by the Energy Market Regulatory Authority (EMRA), approximately 150 companies applied for a license, but only 48 received one. Due to the unfavourable regulatory environment since 2005, 15 of the licensed biorefineries have closed. Among the 33 that remain, only 7 biorefineries actively produce biodiesel. Others have the capacity but have stopped production and locked their doors. After the 2006 legislative changes, biodiesel lost its tax advantages over petroleum-based diesel. For petroleum-based diesel there is no import tariff, only a special consumption tax, whereas for biodiesel there is an import tariff for raw materials, as well as a special consumption tax. Many biorefineries have shut down since 2006, following the introduction of an unfavourable regulatory environment. Currently, there are no obligatory blending requirements for the fuel sector, and a Special Consumption Tax (OTV) is applied to products. The government does not have any incentives in place for new investment or improvement of biodiesel facilities, nor any special loan programmes or loan guarantees (USDA, 2010: 2-3).

At present, the Primary Legislation in place is an Energy Efficiency Law, which came into effect during 2007. Secondary legislation was enforced during 2008, namely the Regulation on Improving Energy Efficiency for the Utilisation of Energy Resources and Energy, the Regulation on Improving Energy Efficiency, UNFC in Transport, and the Regulation on Energy Performance of Buildings. There is also a target of improving energy intensity by 15%, in a cost effective manner, by the year 2020. The government has also passed a Renewable Energy Law, which promotes the use of renewables in a free energy market. The law introduces fee-in tariffs and a purchase obligation for the distribution companies from certified renewable energy producers. The new law allows each new project implemented before 2011 to benefit from seven years of feed-in tariffs. Hydro and geothermal power producers should receive a fixed feed-in tariff of 15% above TETAS's wholesale electricity price, while all other renewable energy producers will receive a tariff 20% above the wholesale electricity price (INOGATE, 2012).

As the USDA (2010: 4) report indicates, the Turkish public and commercial transportation sector uses only diesel for fuel. Turkish regulations do not allow biodiesel producers to sell biodiesel directly to users. Since the producers do not have direct sales licenses, they must sell products to petroleum distribution companies, whose licenses give them authority to sell biodiesel to gas stations or to consumers. Distribution companies must make initial investment in their facilities, to allow for the storage and distribution of biofuels. Currently, there is no legislation obligating distributors to incorporate biofuels within their product lines. Therefore, distributors view biofuels from a purely commercial perspective, and are not willing to invest in biodiesel distribution.

## 6. Renewable Energy Policies and Some Suggestions for Biofuel Energy Promotion

There has been a surge in renewable energy policy-making during the last decade, as countries have sought to respond to energy-related challenges. An indicator of the pace and scale of policy activity is the large number of countries that have implemented renewable electricity targets in recent years. According to the Renewables 2012 Global Status Report, there are currently 73 countries around the world that have implemented policy targets for renewable electricity at the federal or regional levels, up from just 45 in 2005. In 2011, there was a 32 per cent rise in green energy investment worldwide. Investors injected a record \$211 billion into renewables. This represents a 540 per cent rise since 2004 (UNEP, 2012: 2).

Market-oriented measures, which assure a guaranteed price, can be used to promote renewable energies. Green pricing, for instance, enables electricity customers to pay for renewable energies through direct payments on their monthly utility bills (UNEP, 2009: 94). Market prices send strong signals to all sectors to improve energy efficiency and to adopt energy conservation measures. However, it is important to recognise that there can be serious failures in the provision of correct signals from unguided markets.[8] This is mainly because market mechanisms cannot take externalities into account, and the market is inherently short-sighted with a strong tendency to focus on immediate returns at the expense of long-term returns such as sustainable development. As emphasised in a recent study by Erdoğdu and Karaca (forthcoming), policy measures are

[8] For a detailed explanation of this proposition, see particularly Stewart and Gani (1991), and Erdoğdu and Karaca (forthcoming).

the principal tool that governments have at their disposal, and such measures play a crucial role in promoting the domestic industry. Public policy cannot only have a large impact on the health of many industries, but also on their survival. Policy support is required to manage transitional challenges, and supportive government policies have been essential to the development of modern biofuels. Blending regulations, tax incentives, government purchasing policies and other measures are needed to support biofuels.

Policy measures for creating and supporting an industry can be grouped into two categories, direct and indirect measures. Direct measures refer to policies that specifically target the development of an industry. Indirect measures, on the other hand, focus on the creation of an environment suitable for nurturing a nascent industry. Consequently, a range of direct and indirect measures will now be briefly presented and examined.

There is a broad spectrum of potential policies that can be used to achieve renewable energy targets. The main instruments for promoting renewables are feed-in tariffs (FITs), quota obligations, tenders, and (energy) tax exemptions. In most cases, countries decide on one of these instruments and connect it with other political instruments, such as subsidy programmes (sometimes financed by revenue from energy taxes), soft loans, tax allowances, exemptions for renewables from energy taxes, and information campaigns (Reiche and Bechberger, 2004: 846).

According to the latest comprehensive study carried out by the United Nations Environment Programme (2012), there has been vigorous discussion about the comparative merits of different national renewable energy policies over the past two decades. In the late 1990s, there was debate in Europe as to whether countries should adopt FITs or competitive tenders. Countries such as Denmark and Germany used FITs to drive rapid wind energy capacity additions, whereas countries that implemented competitive tenders (e.g., Ireland, the UK, and France) installed only a limited amount of new renewables. France and Ireland switched from competitive tenders to FITs, whereas the UK switched to tradable credits under its new Renewables Obligation. As a result, the most prevalent national renewable energy policy became FIT. As of early 2012, 65 countries had some form of FIT in place, and over half of these are in developing countries (UNEP, 2012: 13).

The UN report outlines some advantages of FITs over other policies, which are as follows:

1. FITs can minimise key investor risks when compared to other policy types, thereby lowering the cost of capital required to finance projects.
2. FIT policies can be designed to target not just on-grid capacity, but also to support off-grid systems (e.g., mini-grids) and, therefore, can help to increase energy access.
3. FITs can streamline the administrative burden and transaction costs of renewable energy policy for both developers and programme administrators. FIT design can influence the degree to which economic benefits from renewable energy development are captured domestically. Hence, when developing FITs, policy makers need to evaluate how FITs interact with existing or proposed policies, taking into account both how FITs might create synergies with other policies and how their interaction may create unintended consequences (UNEP, 2012: 15).

**Table 3-9 Key Policy Instruments in Selected Countries Where E= Electricity, H= Heat, T= Transport, Eth= Ethanol and BD= Biodiesel**

| Country | Policy Instruments | | | | | | | |
|---|---|---|---|---|---|---|---|---|
| | Binding Targets/ Mandates[1] | Voluntary Targets[1] | Direct Incentives[2] | Grants | Feed-in Tariffs | Compulsory grid connection | Sustainability Criteria | Tariffs |
| Brazil | E,T | | T | | | | | Removed |
| China | | E,T[4] | T | E,T | E,H | E,H | | n/a |
| India | T,E[3] | T,BD | E | E,H,T | E | | | n/a |
| Mexico | E[3] | T | E | | | E | | Eth |
| South Africa | T,E | E,T | E,T | | | | | n/a |
| Canada | E,T,H | E[4],T[4] | T | E,H,T | | | | Eth |
| France | | E[3],H[3],T | E,H,T | | E | | | As EU below |
| Germany | E[3],T | | H | H | E | E | E,H,T | As EU below |
| Italy | E[3] | E[3],T | T | E,H,T | E | E | | As EU below |
| Japan | | E,H,T | | | | E | | Eth, B-D |
| Russia | | E,H,T | T | | | | | n/a |
| UK | E[3],T[3] | E[3],T | E,H,T | E,H,T | E | | T | As EU blow |
| USA | T,T[4],E[4] | E[4] | E,H,T | E,T | E | | | Eth |
| EU | E[3],T | E[3],H[3],T | T | E,H,T | | E | T | Eth, B-D |

**Source:** IPCC, 2012, s. 251.

**Notes:** 1. blending or market penetration; 2. fiscal incentives: tax reductions; public finance: loan support/guarantees; 3. target applies to all RE sources; 4. target is set at a sub-national level.

Countries planning to develop domestic biofuel industries will be able to draw important lessons, both positive and negative, from the industry leaders, in particular Brazil, the US, and the EU (WEC, 2010b). Tax incentives have been used successfully in Brazil, Germany, the US and other countries to spur biofuel production and to reduce biofuel prices at the pump. For example, Germany and Austria have implemented tax incentives for pure biodiesel. Hence, biodiesel is 18 per cent cheaper than fossil fuel based diesel in Germany. France and Italy have imposed a 30 per cent tax deduction to promote the use of biodiesel fuels in cities (Soylu et al., 2004). In December 2004, Brazil was estimated to have saved US$ 60.74 billion (valuing gasoline at its price in the international market) due to the substitution of gasoline over the period 1976-200g; taking into consideration interest on the avoided foreign debt, the savings amount to US$ 121.26 billion (24% of the total GDP in 2004). The positive contribution of these policies to the economy can be observed in the decrease in external debt and the security of energy supply. Becker et al. (2011) identified 370 state policies (across all the 50 states in the U.S.) that aimed to provide incentives for forest biomass utilisation. These varied policies were organised by use within the biomass-to-energy supply chain, and were then assessed state-by-state and region-by-region to see how different policies have a positive or negative impact on the supply chain. The Turkish government needs to enact similar policies to those outlined here, in order to encourage investment in biofuels, as implementing similar incentives will lead to an increase in biodiesel use. However, oil seed production in Turkey cannot meet the country's demands, even in today's conditions (Adıyaman and Günay, 2008: 118). Thus, the production of raw materials should be a primary target of these incentives.

According to WEC (2010a: 14), to achieve their full potential in providing security of supply and environmental and social benefits, biofuels need to represent an increasing share of biomass. For this, grandfathering subsidies, tariffs and other tools might be needed until domestic markets have been established. WEC also suggests that all agricultural policies and strategies must be based on local, national or in some cases regional circumstances, and must include a mix of environmental (land, water, climate), social (population, education) and economic (infrastructure, governance) factors. It is therefore impossible to develop a "one-size-fits-all" policy for biofuel production.

In some developed countries, stationary use of biofuels is fostered by feed-in tariffs for the electricity produced. This requires grid connections and available infrastructure. However, if the deployment of feed-in tariffs is not bound to sustainability standards for the biofuels utilised, stationary use may contribute to enhanced environmental pressure and trans-regional problem-shifting. Therefore, when supporting the stationary use of biofuels under a wider sustainability perspective, the overall land use requirements of a country need to be considered in a comprehensive biomass action plan (UNEP, 2009: 92).

Policies have essentially triggered the development of biofuel demand by setting targets and blending quotas. By 2006, mandates for blending biofuels into vehicle fuels had been enacted in at least 36 states/provinces and 17 countries at the national level. Most mandates require blending 10-15 per cent ethanol with gasoline, or blending 2-5 per cent biodiesel with diesel fuel. In addition, recent targets define higher levels of envisaged biofuel use in various countries (UNEP, 2009: 16).

In a number of countries, policies have been established to promote recycling and the energy efficiency of waste management. Policy instruments will need to be further developed to balance potentially competing uses of residues for energy and material use. Some countries have designed specific instruments to foster market entry of renewable based energy technologies, including waste and residues. For instance, in Germany, the Renewable Energy Law 39 guarantees certain feed-in tariffs for power generation from manure and/or certain organic residues from agriculture, which are digested to produce biogas, and also for incineration of forest waste wood, in order to make it competitive with fossil based power generation (UNEP, 2009: 93).

Doornbosch and Steenblick (2007) have suggested that if national governments were to replace biofuel mandates with technology-neutral policies such as a carbon tax, the development of efficient technologies would be better stimulated (UNEP, 2009: 94). A similar policy suggestion would be to shift the tax burden from labour and capital to environmentally related activities, particularly energy and environmental pollution. Altogether, a range of strategies and measures could be used to further develop policies which could contribute effectively to a more efficient and sustainable use of biomass and other resources.

Based on our literature survey, the following modifications to the Turkish renewable energy policies are recommended:

- Ensure that government initiatives relative to biomass supply keep pace with efforts to encourage biomass consumption for energy production.
- Maintain a focus on new sources of supply.

- Credit the distinct sustainability benefits of biomass energy in policy initiatives and/or implement penalties for non-sustainable use of natural gas or other fossil fuel resources.

Another policy that can be recommended is unit energy pricing. Turkey's electricity price does not reflect the contribution of biomass to environmental protection, and hence it is difficult to attract investment capital. The present pricing mechanism for biomass power generation not only lacks market fairness between renewable resources and fossil energy, but also lacks market efficiency between regions with different biomass resource endowments. As Paulos et al. (1998: 6) suggest, another way of encouraging the switch to renewable sources is to fund renewable energy development by the government through a small charge on all electricity sold. Such a charge could fund specific activities to overcome market barriers and help commercialise new technologies. These charges are sometimes referred to as "system benefit charges."

## 7. Conclusion

Facing an increase in its energy consumption, which is above the world average, and being dependent on foreign energy resources, Turkey suffers significantly from energy deficiency and energy insecurity. In addition, the country is increasingly threatened by environmental pollution. Sustainable development requires a sustainable supply of clean and affordable renewable energy sources that do not cause negative societal impacts. Unless it substantially increases the share of renewables in its energy mix, Turkey risks irreversibly damaging both the environment and the natural resource base needed to support economic growth and well-being. Thus, the formidable challenge that Turkey faces today is to transform the economy from being driven primarily by fossil fuel sources of energy to becoming an economy that can function effectively through the use of renewable energy sources. This study has investigated the importance of biomass energy in providing clean and secure energy for Turkey.

The world's transport system is currently based on oil, and demand for oil is expected to grow along with the overall demand for energy. Biofuels can help to meet this demand and, even if they do not completely replace oil, they should be regarded as an integral part of the energy mix. As with all forms of energy production, biomass energy systems raise some environmental issues that must be addressed. In biomass energy projects, issues such as air pollution, impacts on forests, and impacts due to crop cultivation must be reviewed on a case-by-case basis.

This paper has sought to evaluate Turkey's biomass energy potential and the ways by which this potential can be resourced. In these terms, the role of the government was identified as crucial in formulating and implementing favourable policies in order to foster the rapid development of the biomass energy sector in Turkey. It is suggested that achieving sustainable development depends on good governance practices and, in particular, on the effective implementation of national sustainable development strategies that integrate government decision-making within the economic, environmental, and social spheres. To achieve a rapid increase in biofuels production that can be sustained over the long term, policies that are consistent, long term, and supported by broad stakeholder participation are needed.

## References

Adıyaman, A. and Günay, S. (2008), The Importance of Biodiesel for a Solving of High Agriculture Cost Problem in Turkey, Eastern Geographical Review 19, Vol.13, Number 19, pp. 105-122.

Balat, M. (2005), Use of biomass sources for energy in Turkey and a view to biomass potential, Biomass and Bioenergy, Volume 29, Issue 1, pp. 32-41.

Başçetinçelik, A. Karaca, C. Özturk, H.H. Kacıra, M. Kaya, D. Ekinci, K. Barns, I. Baban, A. Nieminen, M. and Komioti, N. (2005), First progress report of exploitation of agricultural residues in Turkey. LIFE 03 TCY/ TR /000061.

Bauen, A., G. Berndes, M. Junginger, M. Londo, F. Vuille, R. Ball, T. Bole, C. Chudziak, A. Faaij, and H. Mozaffarian (2009). Bioenergy: A Sustainable and Reliable Energy Source: A Review of Status and Prospects. IEA Bioenergy: ExCo: 2009: 06, p. 108.

Becker, D., C. Moseley and C. Lee (2011), A Supply Chain Analysis Framework for Assessing State-Level Forest Biomass Utilisation Policies in the United States. Biomass and Bioenergy 35, pp. 1429-1439. Viewed on 21th of October 2012 and retrieved from: http://www.forestry.umn.edu/prod/groups/cfans/@pub/@cfans/@forestry/documents/asset/cfans_asset_313827.pdf

BIOSYNERGY (2008), Advanced Biorefinery Concepts: A Feasibility and Reality Check, Aliki Georgakaki (Ed.), BIOSYNERGY Enlargement and Integration Workshop, Joint Research Centre, Institute for Energy, Petten, the Netherlands.

Bowyer, J. (2011), Potential Impacts of Climate and Energy Policy on Forest Sector Industries: Providing Incentives for Bioenergy While

Protecting Established Biomass-based Industries, June, Dovetail Partners, Inc, Viewed on 21th of October 2012 and retrieved from http://www.dovetailinc.org/files/DovetailBiomassWERCProjectReport .pdf

Cirne, D.G., A. Lehtomäki, L. Björnsson, and L.L. Blackall (2007). Hydrolysis and microbial community analyses in two-stage anaerobic digestion of energy crops. Journal of Applied Microbiology, 103(3), pp. 516-527.

Cremers, M.F.G. (2009). Deliverable 4, Technical Status of Biomass Co-fi ring. 50831165-Consulting 09-1654, IEA Bioenergy Task 32, Arnhem, The Netherlands, p. 43.

Demirbaş, A. (2001), Energy balance, energy sources, energy policy, future developments and energy investments in Turkey, Energy Conandrsion and Management 42, pp. 1239-1258.

—. (2006), Turkey's renewable energy facilities in the near future. Energy Sources, Part A 28, pp. 527-536.

—. (2008), Importance Biomass Energy Sources for Turkey. Energy Policy, 36, pp. 834-832.

—. (2011), Competitive Liquid Biofuels from Biomass, Applied Energy, 88, pp. 17-28.

Dündar, C. and Arıkan, Y. (2004), Enerji, çevre ve sürdürülebilirlik, Bildiriler Kitabı, Türkiye IV. Enerji Sempozyumu, Ankara, 379-390.

Doornbosch, R. and Steenblick, R. (2007), Biofuels: Is the Cure Worse than the Disease?, OECD Round Table on Sustainable Development, Paris, 11-12 September, 2007.

Duygu, A. E. (2004), AB'nin Enerji Bağımsızlığı ve Sürdürülebilir Kalkınma Politikalarında Biyokütle Enerjisinin Yeri: Üye, Aday Ülkelerdeki Araştırmalar ve Uygulamalar. Bildiriler Kitabı, Ankara: Türkiye IV. Enerji Sempozyumu, pp. 451-480.

EC (2008), Energy Sources, Production Costs and Performance of Technologies for Power Generation, Heating and Transport, European Commission, COM(2008)744, Brussels, Belgium.

Econ Poyry (2008). Current Bioenergy Application and Conversion Technologies in the Nordic Countries. 2008-052, Nordic Energy Research, Copenhagen, Denmark, p. 37.

Egsgaard, H.U., P. Hansen, J. Arendt, P. Glarborg, and C. Nielsen (2009). Combustion and gasifi cation technologies. In: Risø Energy Report 2. Risø, Roskilde, Denmark, pp. 35-39.

EPRI (2008), Program on Technology Innovation: Power Generation (Central Station) Technology Options–Executive Summary, Electric Power Research Institute, Palo Alto, United States.

Erdoğdu, E. (2008), An Expose of Bioenergy and its Potential and Utilisation in Turkey, Energy Policy, Vol. 36, No. 6, pp. 2182-2190.

Erdoğdu, M. M. (2012), Sustainable Development Perspective for Dealing with Economic Crises and Combating Unemployment, Labour Markets & Employment, (ed.) D. K. Demirer and K.t Lordoğlu, IJOPEC Publication, London, pp. 111-128.

Erdoğdu, M. M. and C. Karaca (forthcoming), A Road Map for a Domestic Wind Turbine Manufacturing Industry in Turkey.

ETKB (2012), Statistic Data, The Ministry of Energy and Natural Resources, Ankara, Turkey, Viewed on 20th of October 2012 and retrieved from Information Portal: http://www.enerji.gov.tr/index.php?dil=tr&sf=webpages&b=y_istatistik&bn=244&hn=244&id=398

House of the Lords (2008), The Economics of Renewable Energy, 4th Report of Session 2007□08, Vol. I: Report, Select Committee on Economic Affairs, London, United Kingdom.

IEA (2008). Energy Technology Perspectives 2008. Scenarios and Strategies to 2050. International Energy Agency, Paris, France, p. 646.

—. (2009), Energy Policies of IEA Countries Turkey Review 2009.

—. (2010a), Energy Policies of IEA Countries Turkey Review 2009, IEA/OECD 2010.

—. (2010b), Projected Costs of Generating Electricity, International Energy Agency.

—. (2012), Technology Roadmap Bioenergy for Heat and Power, International Energy Agency, OECD/IEA 2012

IEA Bioenergy (2007). Potential Contribution of Bioenergy to the World's Future Energy Demand. IEA Bioenergy: ExCo: 2007: 02, p. 12.

—. (2009). Bioenergy: A Sustainable and Reliable Energy Source. Main Report. IEA Bioenergy: ExCo: 2009: 06, p. 108.

INOGATE (2012), Energy Portal Energy Cooperation between the EU, Eastern Europe, the Caucasus and Central Asia, Viewed on 20th of October 2012 and retrieved from: http://www.inogate.org/index.php?option=com_inogate&view=countrysector&id=37&lang=en

IPCC (2007a). Climate Change 2007: Impacts, Adaptation and Vulnerability. Contribution of Working Group II to the Fourth Assessment Report of the Intergovernmental Panel on Climate Change. M.L. Parry, O.F. Canziani, J. P. Palutikof, P. J. van der Linden, and C.E. Hanson (eds.), Cambridge University Press, p. 979.

—. (2012), Renewable Energy Sources and Climate Change Mitigation, Special Report of the Intergovernmental Panel on Climate Change, Cambridge University Press, 32 Avenue of the Americas, New York.

—. (2007b). Climate Change 2007: Mitigation of Climate Change. Contribution of Working Group III to the Fourth Assessment Report of the Intergovernmental Panel on Climate Change. B. Metz, O.R. Davidson, P. R. Bosch, R. Dave, and L.A. Meyer (eds.), Cambridge University Press, p. 851.

Kaygusuz, K. and S. Keleş (2012), Sustainable Bioenergy Policies in Turkey, Journal of Engineering Research and Applied Science, Vol. 1, pp. 80-91.

Koukouzas, N., A. Katsiadakis, E. Karlopoulos, and E. Kakaras (2008). Co-gasifi cation of solid waste and lignite - A case study for Western Macedonia. Waste Management, 28(7), pp. 1263-1275.

Koyun, A. (2007), Energy Efficiency and Renewable Energy: Turkey–National Study, Plan Bleu Regional Activity Centre, Sophia Antipolis.

Kuuva, K., and L. Ruska (2009). Final Report of Feed-in Tariff Task Force. 59/2009, Ministry of Employment and the Economy of Finland, Helsinki, Finland, p. 101.

Lange, Jean-Paul (2007), Biofuels, Bioproducts and Biorefining, Biofpr, Volume 1, Issue 1, pp. 39-48.

Lettens, P. Muys, B. Ceulemans, R. Moons, E. Garcia, J.and Coppin, P. (2003), Energy budget and greenhouse gas balance evaluation of sustainable coppice systems for electricity production, Biomass Bioenergy, 24, pp. 179-197.

Matthews, W. (2001), Modelling of energy and carbon budgets of wood fuel coppice systems, Biomass Bioenergy, 21, pp. 1-19.

MIT (2009), Update on the Cost of Nuclear Power, Cambridge, United States.

NRC (2009). America's Energy Future: Electricity from Renewable Resources: Status, Prospects, and Impediments. National Research Council, The National Academies Press, Washington, DC, p. 367.

Özertan, G. (2008), Biyoyakıtların Türkiye'nin Enerji, Tarım, Çevre and Kırsal Kalkınma Politikaları İçin Önemi. İktisat İşletme ve Finans, 23 (262), pp. 17-34.

Öztürk, H. H. and Başçetinçelik, A. (2006), Energy Exploitation of Agricultural Biomass Potential in Turkey. Energy Exploration and Exploitation, Volume 24, Number 4+5, pp. 313-330

Paulos, B. Clemmer S. and Nogee, A. (1998), Powerful Solutions for Nebraska Seven Ways to Switch to Renewable Electricity, Union of Concerned Scientist Reports.

Petersson, A. and A. Wellinger (2009). Biogas Upgrading Technologies–Developments and Innovations. IEA Bioenergy Task 37 publication, Malmo, Sweden, p. 20. Viewed on 20th of October 2012 and retrieved from Information Portal: www.biogasmax.com/media/iea_2biogas_upgrading_tech__025919000_1434_30032010.pdf

Reiche, D. and M. Bechberger (2004). Viewpoint: Policy differences in the promotion of renewable energies in the EU member states, Energy Policy, No: 32, pp. 843-849.

Research Reports International (2006), The Use of Biomass for Power Generation in the U.S.

Soylu, P. Karabektaş, M. and Ermiş, K. (2004), Otomobiller için alternatif enerji kaynaklarının incelenmesi. II. Uluslararası Ege Sempozyumu 26-28 Mayıs, pp. 491-504.

Stewart, F. and E. Ghani (1991), How Significant Are Externalities for Development, World Development, Vol. 19 (6), pp. 569-594.

TUIK (2012), Crop Production Statistics, Statistical Data, Viewed on 20th of October 2012 and retrieved from Information Portal: http://www.tuik.gov.tr/VeriBilgi.do?alt_id=45

UN (2007), The United Nations Development Agenda: Development for All, The United Nations Department of Economic and Social Affairs, United Nations Publications: New York.

UNEP (2009), Towards Sustainable Production and Use of Resources: Assessing Biofuels, United Nations Environment Programme.

—. (2012), Feed-in Tariffs as a Policy Instrument for Promoting Renewable Energies and Green Economies in Developing Countries, United Nations Environment Programme.

UNEP-WCMC (2009), The impacts of biofuel production on biodiversity: A review of the current literature, The United Nations Environment Programme and World Conservation Monitoring Centre, December.

USDA (2010), 2010 Turkey Biofuels Annual, USDA Foreign Agricultural Service, 8/13/2010, viewed on 20th of October 2012 and retrieved from: http://gain.fas.usda.gov/Recent%20GAIN%20Publications/Biofuels%20Annual_Ankara_Turkey_8-13-2010.pdf

Vande Walle, I. Van Camp, N. Van de Casteele, L. Andrheyen K. and Lemeur, R. (2007), Short-rotation forestry of birch, maple, poplar and willow in Flanders (Belgium). II. Energy production and $CO_2$ emission reduction potential, Biomass Bioenergy, 31, pp. 276-283.

WEC (2010a) Biofuels: Policies, Standards and Technologies, World Energy Council 2010, United Kingdom.

—. (2010b), 2010 Survey of Energy Resources, World Energy Council, United Kingdom.
—. (2012) World Energy Insight 2012, World Energy Leaders' Summit, Istanbul, Turkey.
Wihersaari, M. (2005), Greenhouse gas emissions from final harandst fuel chip production in Finland. Biomass Bioenergy, 28, pp. 435 43.
Worldwatch (2006), Biofuels for transportation: global potential and implications for sustainable agriculture and energy in the 21st century, Worldwatch Institute, Washington, DC.
Wu, C. Z. and Ma, L.L. (2003), Modern Technologies in Use of Biomass Energies, Beijing: Chemistry Industry Press, pp. 145-178.

# ENERGY AND CLIMATE CHANGE POLICIES OF RUSSIA AND CHINA

ÇAĞLA GÜL YESEVI[1]
AND CEREN UYSAL OĞUZ[2]

## 1. What is Climate Change?

The environment has been the subject of interest in international relations for more than four decades. It is widely accepted that global environmental issues know no boundaries; hence, problems such as pollution, climate change, the depletion of the ozone layer, and excessive consumption of natural resources are of concern to all countries. In order to deal with these matters, international organisations such as the United Nations try to create sets of rules and obligations through summits and conventions. Moreover, environmental issues can generate both conflict and cooperation. Global climate change, which is considered one of the most important environmental problems of the twenty-first century, has the potential to cause international conflict as well as to create opportunities for international cooperation.

In the United Nations Framework Convention on Climate Change (UNFCCC), climate change is defined as "a change of climate which is attributed directly or indirectly to human activity that alters the composition of the global atmosphere and which is in addition to natural climate variability observed over comparable time periods" (United Nations, 1992: 3). As the definition acknowledges, since the beginning of the Industrial Revolution, human activities such as industrial processes, changes in land use, deforestation, agriculture, and transportation have increased enissions of greenhouse gases (GHG) into the atmosphere. Consequently, this has caused significant changes in the global climate

[1] Assistant Professor, Department of International Relations, Faculty of Economics and Administrative Sciences, Istanbul Kültür University, TURKEY
[2] Assistant Professor, Department of International Relations, Faculty of Economics and Administrative Sciences, Istanbul Kültür University, TURKEY

system, affecting water resources and food supplies, as well as creating political, social, economic, and ecological problems.

Climate change may be a source of instability and potential conflict between states as a result of competition concerning access to, and control of, energy resources. For example, the Climate Change and International Security Report (2008), prepared by the European Commission and Javier Solana, and presented to the European Council, states that "because much of the world's hydrocarbon reserves are in regions vulnerable to the impacts of climate change and because many oil and gas producing states already face significant social economic and demographic challenges, instability is likely to increase." International cooperation and solidarity should deal with the security challenges created by climate change. However, most of the time, prioritising national economic and political interests over environmental concerns gets in the way of such cooperation.

The Russian Federation and China are both key actors in terms of global energy supply and GHG emissions. For this reason, their active participation is indispensable in an effective climate change mitigation and adaptation structure, especially in the post-Kyoto period. Therefore, it is necessary to examine the energy and climate change policies of these countries, in order to evaluate their contribution to the international climate regime and the importance of their energy strategies in international relations.

## 2. Russian Energy Strategy

Russia is one of the world's largest energy suppliers. In glabal terms, it possesses 33% of natural gas resources, 10% of oil reserves, 20% of coal reserves, and 14% of uranium reserves. The aim of its energy policy is to ensure the effective use of its energy resources. The 2020 Energy Strategy of Russia, issued by the Ministry of Energy, reveals a liberal understanding of the issue. Firstly, it established the goal of reaching a new and improved quality of fuel and energy complex. Secondly, it decided to focus upon the growth of competitiveness and the openness of production and services on the global market. The Ministry indicated that the role of the state should be limited to forming market infrastructure and to working as a regulator of market relationships (The Ministry of Energy of the Russian Federation, 2003: 2-3).

It is clear that Russian oil and natural gas industries have played a significant role on the world energy market. President Putin believes that Russia's natural resources will not only secure the country's economic development, but also serve as the guarantor of the country's international position (Olcott, 2004: 16). Although the 2020 Energy Strategy presented a liberal perspective, the Russian Federation previously employed an authoritarian system, with the government largely owning these industries and making decisions through a top-down process. Moreover, there has been a tendency to control the pipelines, ports, and storage facilities of Central and Eastern Europe and former Soviet Republics. This policy has led to concern in Central and Eastern European states that Russia could take advantage of this inherent energy dependency to interfere in the domestic affairs of these states, and force them to make foreign policy concessions. Furthermore, these states were also worried that the Russian energy companies, through the control of the regional energy infrastructure, might favour some local businessmen and politicians and seek to influence the internal political situation (Woehrel, 2009: 1-2).

Former President Mikhail Gorbachev lists some of the problems of the Russian Federation (Dünya Gündemi, 2010), and one worth mentioning is that the Russian economy was based primarily on the energy sector at the time when democratic and liberal rights were suppressed. This economic dependence on energy brought many problems, due to the fact that these resources are limited. Poor technological conditions, difficulties in drilling, and an authoritarian administration harmed the exploration, production and exportation of natural resources. Moreover, terrorist activities, ethnic conflicts, difficulties in payment, and transportation and natural disasters all influence the Russian energy policy and therefore its economy. Furthermore, poor weather conditions affecting transport and the changing prices of natural resources may hinder the development of the Russian economy. It is apparent that any negative impact on the economy could influence and change Russian political and social life for the worse. The collapse of the Soviet Union saw the establishment of oligarchies and the privatising of the energy sector. However, when Putin came to power, the administration changed, and he initiated a patriotic and authoritarian rule, waging war against the oligarchs of Yeltsin's time.

One major issue relating to the Russian energy sector was the arrest of Yukos CEO Mikhail Khodorkovsky, on 25 October 2003. Yukos was one of the big oil companies of the Russian Federation (Olcott, 2004: 1), producing 20% of Russian oil until 2003. Khodorkovsky has recently been charged with stealing all the oil his company ever produced, and in 2005 he was sentenced for failure to pay taxes on total oil production (Ioffe, 2010). Yukos' assets were frozen and, as a result, the tax debt could not be paid (BBC News, 2005). Vladimir Putin has stated that private property and a free market economy are important, but that the rights of Russian citizens do not take precedence over the national interests of the Russian

Federation. Putin described this policy as "managed democracy" and stated that the premature globalisation of the Russian economy would lead to greater hardship for the majority of Russian people. As a result, it was deemed that wealth would be concentrated among only a few people (Olcott, 2004: 3-4).

Another company, Gazprom, founded in 1993, has identified itself as the world's largest energy business engaged in natural gas, gas condensate and oil prospecting, transmission, processing and marketing, as well as power generation both inside and outside Russia. Gazprom accounts for a 78% share of total Russian gas production, and also produces 15% of global gas production. The Russian state holds 50% of its shares. It employs 404.4 thousand people. The strategic goal of Gazprom is to become a leader among global energy companies by conquering new markets, diversifying business activities, and pursuing supply security (gazpromquestions.ru). It has been alleged that Gazprom's key current gas fields are in decline, that its infrastructure is ageing, and that the industry faces investment challenges. Some foreign experts believe that access to foreign investment in providing expertise and capital is limited (Woehrel, 2009: 2). The market value of Gazprom declined from 300 billion dollars to 130 billion dollars within two and a half years. Moreover, the German energy firm EON has decided to sell 3.5% of its Gazprom shares (Dünya Gündemi, 2011: 7).

Rosneft is an oil company mostly owned by the Russian state, and Transneft is a wholly state-owned oil pipeline company transporting most of Russia's oil. Lukoil is a prominent privately owned energy company in the Russian Federation, ensuring 18.6% of Russian oil production and 18.9% of total Russian oil refining. It produces 2.4% of global oil production (lukoil.com).

Financially, the Russian Federation has increased its wealth, but heavy industry, the transport sector, and social infrastructure have all deteriorated. As a result, structural and administrative reforms have not been realised. It is also a fact that domestic gas production is declining, and Gazprom is increasingly reliant on Central Asian gas imports. Nationalising resources in the energy sector would impair oil and gas production, as foreign investment is vital in modernising these sectors. The lack of expertise working in oil and gas offshore is a prime example of where foreign investment could help (Milov, 2008: 9-11). Nationalising resources was defined as a reaction to the emergent capitalism of the 1990s. Boris Yeltsin opened energy resources to western firms (Helm, 2006) which had until then (through the Patrimonial Muscovite system) been under state control, with ownership of energy resources having been the main parameter of Russian energy policy (Blank, 2006: 16).

The proposed South Stream gas pipelines will transport Russian gas to Western Europe, bypassing Ukraine. The pipes will run from Russia, under the Black Sea to the Bulgarian coast (Engdahl, 2010), and include the Italian ENI. The Russian South Stream project is supported by the German hydrocarbon subsidiary of the powerhouse BASF, and the French Electricity group EDF is also expected to participate. Gazprom's chief, Alexei Miller, has mentioned that this deal means that the South Stream project had been established, and that Gazprom will remain a supplier of gas to Europe for many decades to come. He adds that Gazprom had the markets and the volume of natural gas (The Local, 2011).

The Russian administration is aware of the fact that the fossil fuel sector is vital to its economy and political reach. Through its oil and natural gas production and exportation, Russia maintains its economic development; at the same time, by using this dominant position as a foreign policy tool, it is becoming an important player both regionally and globally.

According to the Energy Strategy of Russia for the Period up to 2030, issued by Ministry of Energy of the Russian Federation in 2010, Russia provides 12% of the world oil trade. Russia holds 23% of the world natural gas reserves and provides 25% of the world trade in natural gas. Additionally, Russia holds 19% of the world's coal reserves.

It is stated that within the period up to 2030, the export of energy resources will remain the major development factor for the Russian economy. It is projected that the impact of the energy sector on the economy will decrease within this period. The main markets for Russia will be Europe and the Commonwealth of Independent States. The assumption is that, looking forward, the proportion of European energy consumption in relation to the total volume of Russian energy export will decline, due to export diversification to Eastern energy markets.

Energy security is one of the most important components of national security. The main problems in this field are the high degree of fixed assets depreciating in the fuel and energy complex, and the low level of investment therein. This is exacerbated by Russia's economic dependence on its energy sector, its failure to match international scientific and technical levels in terms of environmental standards, and the speed of development of the energy infrastructure in Eastern Siberia and the Far East. The strategic objective of the foreign energy policy is the maximum and efficient use of Russia's energy potential for full scale integration into the world energy market, in order to secure the highest possible profit for the national economy.

## 3. Russian Climate Change Policy

Russia is one of the world's major producers of carbon emissions and the largest international sink, due to its vast forests and wildlife areas; it thus plays an important role in the global climate system (Mirovitskaya 2004: 349). This argument was proved correct when the Kyoto Protocol entered into force on February 16, 2005, subsequent to the ratification of the Russian Federation. In Article 25 of the Protocol, it is stated as a prerequisite for entry that no less than fifty-five signatories to the UNFCCC (including developed countries, which account for at least 55% of total carbon dioxide emissions in 1990) should ratify, accept, approve or accede to it. Given that the United States, the largest emitter of the base year, withdrew from the Kyoto negotiations in 2001, Russia with its 16.4% share of the 1990 emissions, was the only country that could prevent the failure of the Protocol (Lioubimtseva, 2010: 2).

Russia signed the UNFCCC at the Rio Summit in 1992 and ratified it in 1994. Also in 1994, the Interagency Commission on Climate Change Issues was established by the government, in order to coordinate national activities outlined in an action plan for domestic implementation of obligations under the Convention, and to evaluate the national position regarding global climate regime (Kotov and Nikitina, 1995: 25; Mirovitskaya, 2004: 350). However, Russia remained reluctant to ratify the Kyoto Protocol. For instance, at the beginning of December 2003, one of the top economic advisers to Putin declared that Russia would withdraw from the Kyoto process, using almost the same argument as the then US president G. W. Bush, namely that: "it was not in the country's economic interest" (Gelbspan, 2004: 11; Myers and Revkin, 2003). As a result, the Kyoto Protocol's ratification process was delayed while Russian policy-makers openly demanded benefits. Consequently, after gaining the EU's support for accession to the World Trade Organisation, along with the prestige of being the principal facilitator of the Protocol (Korppoo and Moe, 2007: 2). Russia ratified the Kyoto Protocol in November 2004.

In the light of the ratification process, Korppoo and Moe (2007: 2) explain the Russian approach to international climate politics as follows:

> …direct benefits related to climate change pacts play a key role encouraging Russia to cooperate. One of the explaining factors is the fact that climate change is still regarded by many as not being a serious environmental problem. Russia has more immediate environmental

problems on its territory than those posed by climate change, and it is not uncommonly argued that climate change could even benefit the country.

According to Korppoo (2008: 2), "the Kyoto Protocol was never seen as an environmental pact in Russia, but rather as a means of gaining economic and political benefits". Moreover, many important Russian scientists working on the subject are sceptics, arguing that there is no serious threat to the climate resulting from human activities. Public opinion on the effects of climate change is not totally negative, with some claiming that higher temperatures would be advantageous, especially with respect to energy consumption and agriculture (Korppoo and Moe, 2007: 2; Korppoo, 2008: 5; Lioubimtseva, 2010: 8).

Therefore, it is suggested that the incentive for the Russian authorities to take part in the Kyoto process is the expectation of economic and political gain. Russia's commitment under the Kyoto Protocol is to maintain greenhouse gas emissions at the 1990 level. When Russia ratified the Protocol in 2004, their emissions were already 33% below the base year level, and the fulfilment of the Kyoto target in the 2008-2012 commitment period seemed to be easily achieved without additional domestic action. Furthermore, as a result of inefficient energy production, transportation, and the consumption infrastructure of the Soviet era, Russia has a huge potential to attract Joint Implementation (JI) projects (Korppoo and Moe, 2007: 3-4; Korppoo, 2008: 3).

Russia is considered one of the key actors in the post-2012 negotiations, since it is the world's third largest emitter of greenhouse gases after China and the USA, with 5.9% of total emissions (Korppoo and Moe, 2007: 7; Lioubimtseva, 2010: 1). According to Korppoo (2008: 7), in order to be convinced to participate in the post-Kyoto climate regime, Russia needs to be recognised as an international actor, which may be a way of re-establishing the country's status and prestige as a world power. In their preparations for the meeting in Copenhagen in 2009, the Russian government declared that their aim was to reduce national emissions by 25% of the base year (1990) by 2020. However, that offer was revised to 10-15% when the Copenhagen Accord failed to set legally binding emission reduction targets (Lioubimtseva, 2010: 1). During the Copenhagen meeting, Russia also emphasised that all major emitters, such as the US and China, should participate in the post-2012 agreement. Afterwards, in line with the unsuccessful negotiations for the post-Kyoto regime in Durban in 2011, Russia (along with Japan and Canada) declared its opposition to a second commitment period under the Kyoto Protocol (Andonova and Alexieva, 2012: 615-617).

For Russia, focusing on concrete policies such as energy efficiency could have a positive impact on the economy, while at the same time reducing emissions (Korppoo, 2008: 4). For instance, former President Dmitry Medvedev, in his opening speech at the Meeting on Improving Environmental and Energy Efficiency in the Russian Economy in June 2008, mentioned the importance of the development of environmentally sound technologies and energy efficiency in buildings, structures and production processes. He also emphasised the need to fund "projects involving the use of renewable energy and the introduction of environmentally-friendly and energy efficient technologies" (Medvedev, 2008).

The role of fossil fuel companies is the key to the country's economic structure, and the Russian administration's influence on both public and private companies is a crucial element in this respect. If they take part in the formulation process of the national climate change policy by investing in clean technologies and efficiency improvement, Russia could agree to higher emission reduction targets, while at the same time securing its economic development goals.

## 4. Chinese Energy Policy

The Chinese economy has grown rapidly across the industrial sector, and it is a fact that Chinese economic growth has resulted in new demands on world energy supplies. From 1953 to 1978, Chinese industrial development depended on a few basic industrial materials, metals and machinery. However, the technological capacity and output of Chinese machinery was very low. At that time, the Chinese economy was capital and energy-intensive. After 1978, China began to produce low-technology, labour-intensive consumer goods, and these sectors ensured excellent opportunities for the Chinese economy. Nonetheless, light and diversified manufacture ended after 1995, after which higher-technology industries emerged. Electronics and telecommunication equipment increased their share, according to the 2004 census (Naughton, 2007: 329-332).

There are two important aspects of energy policy—energy security and global climate change. The International Energy Agency (IEA) (2010) suggested that global demand for fossil fuels such as coal, natural gas and oil would increase, and that the largest energy demands would come from China and India. After the financial crisis of 2008-2009, China became the world's largest energy user, overtaking the US. The IEA has claimed that not only will the global demand for natural gas increase, but that the Chinese demand for gas will grow even faster. They also stated that the demand for electricity will rise, expecting it to triple between 2008 and 2035. Coal and natural gas will remain the main sources for producing electricity until 2035, even as China emerges as a leader in wind power. Additionally, the IEA indicated that the use of biofuels will be supported, and that their share will increase four-fold by 2035. Biofuels are transport fuels derived from biomass feedstock. At a global level, this sector of energy will develop, especially in the US, the EU and Brazil. The IEA suggested that the resources of the Caspian Basin should be developed (IEA, 2010: 45-52).

China has surpassed Japan to become the world's second largest economy (Dünya Gündemi, 2011), the number-one consumer of energy (Nicholson, 2011: 1), and the number-three producer of energy after the US and Russia. The Chinese energy sector has depended almost exclusively on coal, resulting in many environmental problems. Small-scale mines are profitable, but they are dangerous and dirty. China has been trying to use energy resources more efficiently, and economic reforms on energy conservation have been implemented, including the modernisation of the technologies employed. While Chinese industry remains the biggest user of energy, the Chinese people have also begun to consume more (Naughton, 2007: 334-339).

China is now the second largest oil importer in the world, with nearly one third of global oil demand. Moreover, the consumption of aluminium, copper, nickel and iron has doubled (Zweig and Jianhai, 2005: 25). Due to political unrest in North Africa and the Middle East, from which China has been importing 60% of its oil (Naughton, 2007: 342), the country has renewed focus on energy conservation. As a net importer of oil, China's energy needs coincide with its national security. In order to reduce its dependence on foreign oil and to boost its security, China has been trying to produce electric cars (Bradsher, 2011: 1-2). Add this to the fact that China is the largest importer of coal in the world (Revkin, 2011: 2), and the need for energy conservation becomes apparent.

The Chinese economy is based on a state-owned energy sector. The Chinese government controls electricity distribution systems, oil, gas and electricity companies, as well as many coal-mining companies. A key trend is that the Chinese economy has been growing faster than government forecasts have predicted. In 2009, the government decided to reduce its energy usage by 20%, but could not achieve this goal. As a result, an "iron-hand campaign" began, and town officials began to shut off the electricity supply to businesses, homes, and hospitals. The government ordered the closure of two thousand factories. This resulted in a 19% improvement in conservation, realised in January 2011 (Bradsher, 2011: 3-4).

The major players of the Chinese energy sector are PetroChina, Sinopec and Cnooc. They operate in thirty-one countries, including Angola, as well as in more dangerous areas such as Sudan, Venezuela and Iran, and, more recently, Libya. In 1993, China was the most important oil exporter, but domestic production stagnated and China became a net oil importer (Nicholson, 2011: 1-2).

In the Tenth Five Year plan of 2001, a speech by Prime Minister Zhu Rongji revealed the principles of the Chinese energy strategy, the main parameters being the diversification of the sources of energy imports, enhancing overseas investments by state oil companies, increasing investment in oil and natural gas, opening more import channels, establishing new government controlled petroleum reserves, and reducing dependence on oil (Blank, 2006: 12).

The Information Office of the State Council, China's cabinet, published a white paper on the country's energy policy in 2012. In the preface, it is stated that China adopted a policy of reform, and that by opening up in the late 1970s, its energy industry made great advances. It is noted that China has been declared the world's largest energy producer and has built up a comprehensive energy supply system comprising coal, electricity, petroleum, natural gas, and new and renewable energy resources. Moreover, its universal energy service and civil energy use conditions have markedly improved. It also highlights the fact that China's energy sector faces many challenges. Some of the problems facing the Chinese energy sector are high energy resource endowment, and a low per-capita share of coal, petroleum, and natural gas. It is obvious that China's energy consumption has been growing too quickly. Fossil energy resources have been exploited on a large scale, causing a certain amount of damage to the eco-environment, but China has been trying to curb excessive consumption of energy resources and attempting to achieve a comprehensive, balanced, and sustainable development of the economy, society, and ecology. In this sense, China aims to strengthen its efforts in energy conservation and emission reduction, and strives to raise the efficiency of energy utilisation. The 2012 white paper states that China relies on scientific, technological, and system innovation to raise efficiency in all aspects of energy utilisation, to further develop new and renewable energy resources, and to promote the clean and efficient development and utilisation of fossil energy resources.

In Part IV of the 2012 white paper, it is proposed that China should vigorously develop new and renewable energy. By the end of the Twelfth

Five Year Plan, China has endeavoured to increase the shares of non-fossil fuels in primary energy consumption and to install a generating capacity of 11.4% and 30%. To that end, China will develop its hydropower, nuclear power, wind power, solar, biomass, and other types of renewable energy. The 2012 white paper goes on to state that China will promote distributed utilisation of clean energy and will try to accelerate the construction of distributed energy systems (DES) for natural gas at energy load centres. Furthermore, China will spread the technology of distributed renewable energy, especially in high energy-consumption centres, including cities and industrial parks. In addition, China aims to carry out distributed renewable energy construction in rural areas, forest regions, and islands, in accordance with local conditions.

The 2012 white paper concludes that energy has been a vital material base for China to modernise and to build a moderately prosperous society. In that sense, the Chinese government strives to address the energy problem by following the sustainable road of energy development. It is projected that China will continue to accelerate in terms of industrialisation and urbanisation for a long time to come. It faces the challenging tasks of developing its economy and improving its people's livelihoods. It is thus presumed that Chinese energy needs will increase in the future. China must rely on itself to increase the energy supply steadily, in order to satisfy energy demands. As stated in the 2012 document, China will try to secure its energy supply with international cooperation and engage with other countries and international energy organisations, working together with them to promote sustainable energy development around the world.

## 5. Chinese Climate Change Policy

China signed the UNFCCC in June 1992 in Rio and ratified it in December 1992. It signed the Kyoto Protocol in May 1998, which it ratified in August 2002. According to the Kyoto Protocol, China, as a developing country, does not have any emission reduction targets. Its obligations are to report its emissions via a national communication, and to outline the steps that are being taken or planned to implement the Convention (Oberheitmann and Sternfeld, 2009: 139).

China is amongst the world's largest greenhousee gas emitters along with the US, and its emissions have increased by almost 80% since 1990 as a result of rapid economic growth and growing energy demand (Pew Center on Global Climate Change, 2007: 1). By 2030, China is expected to be responsible for almost half of all global $CO_2$ emissions (Gilley, 2012:

289). On the other hand, it has the potential to reduce emissions through sustainable policy choices. For instance, the Chinese economy grew by 36% between 1995 and 2000, and owing to their investment in renewable sources, their energy efficiency, and their conservation policies, they were able to cut their emissions by 19% during the same period (Gelbspan, 2004: 38, 72).

As a non-Annex I country, China is able to participate in the Clean Development Mechanism (CDM) projects (Oberheitmann and Sternfeld, 2009: 140). According to "China's Policies and Actions for Addressing Climate Change: The Progress Report," prepared by the Chinese National Development and Reform Commission (2009: 39), China approved 2,174 CDM projects up to June 2009, and companies have been developing such projects to promote renewable energy technologies and energy conservation and to lower energy consumption, thus reducing GHG emissions.

In its 2007 National Climate Change Programme, while emphasising the importance of international cooperation in dealing with climate change, China also mentions that developed countries should fulfil their commitments to providing financial assistance and to transferring technology to developing countries, in order to help them increase their capacity (People's Republic of China National Development and Reform Commission, 2007: 58).

President Hu Jintao declared that China would combine its economic and social development plans and actions on climate change. First of all, efforts to improve energy conservation and energy efficiency will be intensified. The aim is to reduce the 2005 $CO_2$ levels by 2020. Another important goal is to develop renewable energy and nuclear energy, thereby increasing the share of non-fossil fuels in energy consumption by 15% by 2020. Increasing the country's carbon sinks through forestation (to coverage of forty million hectares) is also projected. China is committed to developing a green, low-carbon economy, and to improving research, development and distribution of climate-friendly technologies (People's Republic of China National Development and Reform Commission, 2009: 4).

With its dramatically increasing GHG emissions and mounting energy demand, China will be a key actor in future climate negotiations. Being a developing country creates a dilemma when it comes to climate change mitigation and adaptation. Development goals seem to be in conflict with environmental concerns. China is trying to overcome this dilemma by prioritising energy efficiency and investing in renewable sources, while urging international cooperation and the fulfilment of developed countries' responsibilities.

## 6. Conclusion

Global and national energy policies are multidimensional; they include political, economic, strategic, technological, geographical, and environmental aspects. In this chapter, we have analysed the interconnected energy and climate change policies of two important actors, Russia and China.

First of all, in terms of international relations, there has been a widespread belief that wars and conflicts occur in order to acquire energy resources. It has been alleged that the motivation for the occupation of Afghanistan was to build a pipeline through the country which would take natural gas from Turkmenistan and deliver it to the energy demanding regions of the world. This view was supported by the US Assistant Secretary of State for South and Central Asian Affairs, Richard Boucher; he stated that one of their goals was to stabilise Afghanistan so that it could become a conduit and hub between South and Central Asia, allowing energy could flow to the south. However, a Turkmen official explained that they had no urgent need of the pipeline that the US and the EU had been advancing. Subsequently, Russia and Turkmenistan agreed to build an east-west pipeline, connecting all of the gas fields in one network. China and Turkmenistan also agreed to open a gas pipeline (Ismi, 2010). In relation to the Russian-Chinese relationship, the construction and control of pipelines have been the main issues. Moreover, Russia does not want China or any other state to have any direct involvement in the energy resources of Russia and Central Asia, as Transneft and Gazprom have pipeline monopolies (Blank, 2006: 25). This explains why fossil fuel politics are at the centre of international relations.

In terms of domestic policies, both China and Russia have been trying to update the low-technology structure of their energy sector through a modernisation process. Both states aim to ensure energy efficiency. Therefore, technological changes and structural and institutional reforms have been applied. The opening of energy sectors was considered a challenge to nationalised resources, leading, briefly, to a dilemma between the realisation of energy efficiency and foreign investment and resource nationalism. The authoritarian policies of China and Russia would be in contradiction with these liberal purposes. Although we have witnessed globalisation with multinational firms, it should be taken into consideration that the Russian and the Chinese energy sectors have been developed through nationalist strategies.

In terms of environmental concerns, there is an indisputable link between energy policies and the fight against climate change. Fossil fuels (coal, oil, and natural gas) constitute the largest percentage of the global energy supply. Increasing scientific evidence shows that mitigating and adapting to climate change requires a major shift in current energy policies. Therefore, both developed and developing countries should invest in renewable energy sources such as wind, solar, wave, hydropower, and biodiesels. Technology regarding these alternatives should be shared with developing countries, allowing them to reduce their dependence on fossil fuels, as well as their GHG emissions. On the other hand, the predicted dominance of fossil fuels in the coming decades makes scientific research and technological development necessary in terms of clean energy. So, as mentioned by the Zero Emission Fossil Fuel Power Plants Technology Platform (ZEP), the EU's experience of $CO_2$ separation and underground storage could be applied to power plants, while international research and development cooperation on this technology, particularly with China and Russia, should be encouraged, thus enabling these countries to achieve the required emission reductions (European Commission, 2006: 8-9).

In conclusion, the energy policies of countries like Russia and China should be coordinated with their climate change policies, considering their shares in global energy supply and demand. Investing in renewable sources, improving energy efficiency, and cooperating with the European Union in research and development efforts regarding clean technologies would help these countries play a more significant role in international climate negotiations, while making them effective global actors. Mitigation and adaptation to climate change should become an integral part of these countries' national and foreign policies. Reducing their dependency on fossil fuels would also allow them to pursue more stable relations in a turbulent geo-political climate. Russia and China are indispensable countries in the development of an international climate change regime, which should be based on cooperation between developed and developing countries.

## References

Andonova, Liliana B. and Alexieva Assia (2012). "Continuity and change in Russia's climate negotiations position and strategy". Climate Policy 12 (5), pp. 614-629.

BBC News (2005). "The Timeline: The Rise and Fall of Yukos". http://news.bbc.co.uk/2/hi/business/4041551.stm, (accessed March 15, 2011).

Blank, Stephen (2006). Russo-Chinese Energy Relations: Politics in Command. GMB Publishing Ltd., London.

Bradsher, Keith (2011). "China Reportedly Plans Strict Goals to Save Energy", The New York Times, March 4.

Climate Change and International Security (2008), Paper from the High Representative and the European Commission to the European Council, http://www.consilium.europa.eu/uedocs/cms_data/docs/pressdata/en/reports/99387.pdf, (accessed March 13, 2011).

Dünya Gündemi (2010). "Rusya Gerçek ve Etkili Bir Demokrasiye İhtiyaç Duymakta", December 19–December 26, 314.

Dünya Gündemi (2011). "Alexei Miller: Müşterilerimizin sadece doğalgaz talebini karşılama hizmeti vermekteyiz", January 16–January 23, 317.

Dünya Gündemi (2011). February 27–March 6, 324.

Engdahl, William, F. (2010). "Pipeline Geopolitics: The Russia German Nord Stream Strategic Gas Pipeline", Global Research http://www.globalresearch.ca/index.php?context=va&aid=20080 (accessed March 28, 2011).

European Commission (2006). A Vision for Zero Emission Fossil Fuel Power Plants, Luxembourg.

Gazprom Website, http://eng.gazpromquestions.ru, (November 29, 2012).

Gelbspan, Ross (2004). Boiling Point. Basic Books, New York.

Gilley, Bruce (2012). "Authoritarian environmentalism and China's response to climate change", Environmental Politics 21 (2): 287-307.

Helm, Dieter (2006). "Russia's Energy policy: Politics or Economics?" Open Democracy, October 16, http://www.opendemocracy.net/globalization-institutions_government/russia_energy_4004.jsp, (accessed March 18, 2011).

Ioffe, Julia (2010). "The Verdict Is In". Foreign Policy, December 30, http://www.foreignpolicy.com/articles/2010/12/30/the_verdict_is_in, (March 15, 2011).

Ismi, Azad (2010). "U.S. and its Allies Foiled: Russia, China, Iran defeat the US in the pipeline wars". Canadian Centre for Policy Alternatives, http://www.policyalternatives.ca/publications/monitor/us-and-its-allies-foiled, (March 28, 2011).

Korppoo, Anna (2008). Russia and the Post-2012 Climate Regime: Foreign Rather than Environmental Policy, The Finnish Institute of International Affairs Briefing Paper No.3, http://www.upi-fiia.fi/assets/events/UPI_Briefing_Paper_23_2008.pdf, (accessed March 20, 2011).

Korppoo, Anna and Moe, Arild (2007). Russian Climate Politics: Light at the End of the Tunnel? Climate Strategies Briefing Paper, April 2007, http://www.climatestrategies.org/research/our-reports/category/79/193.html, (March 20, 2011).

Kotov, Vladimir and Nikitina, Elena (1995). "Russia and International Environmental Co–operation", Green Globe Yearbook of International Co-operation on Environment and Development 1995, Helge Ole Bergesen, Georg Parmann, and Øystein B. Thommessen (eds.), Oxford: Oxford University Press, pp. 17-27. http://www.fni.no/ybiced/95_01_kotov.pdf, (accessed March 20, 2011).

Lioubimtseva, Elena (2010). "Russia's Role in the Post-2012 Climate Change Policy: Key Contradictions and Uncertainties", Forum on Public Policy Online, 2010, (3). http://www.forumonpublicpolicy.com/spring2010.vol2010/spring2010 archive/Lioubimtseva.pdf, (accessed March 20, 2011).

Lukoil Website, http://www.lukoil.com, (March 18, 2011).

Medvedev, Dmitry (2008). Opening Remarks at the Meeting on Improving Environmental and Energy Efficiency in the Russian Economy, June 3, 2008, Moscow, http://archive.kremlin.ru/eng/speeches/2008/06/03/2225_type82912type82913_202070.shtml, (accessed March 22, 2011).

Milov, Vladimir (2008). Russia and the West: The Energy Factor. Center for Strategic and International Studies, Institut Français des Relation Internationales, http://csis.org/files/media/csis/pubs/080731_milov_russia&west_web.pdf, (March 21, 2011).

Mirovitskaya, Natalia (2004). "Trump Card in Kyoto Pact: Russian Federation's Interests and Positions on the Global Climate Change Regime", Climate Change Five Years after Kyoto (ed. Velma I. Grover), Science Publishers, Enfield, pp. 349-376.

Myers, Steven Lee and Revkin, Andrew C. (2003). "Russia to Reject Pact on Climate, Putin Aide Says". The New York Times, December 2, A12.

Naughton, Barry (2007). The Chinese Economy: Transitions and Growth, MIT Press, Cambridge.

Nicholson, Chris (2011). "China's Energy Industry Pushes into Developed Markets". The New York Times, March 14.

Oberheitmann, Andreas and Sternfeld, Eva (2009). "Climate Change in China: The Development of China's Climate Policy and Its Integration into a New International Post-Kyoto Climate Regime", Journal of Current Chinese Affairs, 38 (3), pp. 135-164.

Olcott, Martha Brill (2004). The Energy Dimension in Russian Global Strategy: Vladimir Putin and The Geopolitics of Oil. The James A. Baker III Institute for Public Policy of Rice University, http://www.rice.edu/energy/publications/docs/PEC_Olcott_10_2004.pdf, (accessed March 15, 2011).

People's Republic of China National Development and Reform Commission (2007). China's National Climate Change Programme, June, http://www.ccchina.gov.cn/WebSite/CCChina/UpFile/File188.pdf, (accessed March 24, 2011).

—. (2009). China's Policies and Actions for Addressing Climate Change: The Progress Report, November. http://www.ccchina.gov.cn/WebSite/CCChina/UpFile/File571.pdf, (accessed March 24, 2011).

Pew Center on Global Climate Change (2007).Climate Change Mitigation Measures in the People's Republic of China, International Brief No.1, April. http://www.pewclimate.org/docUploads/International%20Brief%20-%20China.pdf, (accessed March 24, 2011).

Revkin, Andrew (2011). "Behind China's Shift on Energy and Growth". The New York Times, March 5.

The International Energy Agency (2010). World Energy Outlook 2010, Paris.

The Local (2011), "BASF Boosts Russian Gas Pipeline", March 22. http://www.thelocal.de/money/20110322-33873.html, (accessed March 19, 2011).

The Information Office of the State Council, China's cabinet (2012). The 2012 edition of White Paper: China's Energy Policy.

The Ministry of Energy of the Russian Federation (2003). The Summary of the Energy Strategy of Russia for the Period of up to 2020, Moscow, http://ec.europa.eu/energy/russia/events/doc/2003_strategy_2020_en.pdf, (accessed March 19, 2011).

The Ministry of Energy of the Russian Federation (2010). Energy Strategy of Russia for the Period up to 2030, Moscow.

UNFCCC (2005).Caring for climate: A guide to the Climate Change Convention and the Kyoto Protocol, Bonn, http://unfccc.int/resource/docs/publications/caring2005_en.pdf, (accessed March 22, 2011).

United Nations (1992).United Nations Framework Convention on Climate Change. http://unfccc.int/resource/docs/convkp/conveng.pdf, (accessed February 6, 2011).

Woehrel, Steven (2009). Russian Energy Policy toward Neighboring Countries. Congressional Research Service, 2 September, http://www.fas.org/sgp/crs/row/RL34261.pdf, (accessed February 18, 2011).

Zweig, David and Jianhai, Bi (2005). "China's Global Hunt for Energy", Foreign Affairs 84 (5).

# The Relationship between Natural Resources and Security (with an Emphasis on Energy Security)

Hakan Korhan[1]

The insufficiency of available global energy resources in meeting current and future needs constitutes the subject of a literature which keeps on accumulating. The factors that render the subject more significant, in terms of international relations, are that in order to meet increasing energy demands, the dependency on oil, natural gas and coal (also referred to as fossil resources) continues and will continue into the foreseeable future, and that such fossil resources are concentrated in certain regions and/or countries. In other words, the dependency on certain countries and/or regions in supplying the quantities used to meet increasing energy demands is increasing rapidly. Increasing dependency (along with the economic and political consequences of this dependency) has begun to be perceived as a national security problem. In this regard, starting from the 1990s, energy security has been an increasinglty central concept in security definitions.

From this point of view, the place of energy security in national security definitions will constitute the conceptual framework of this study, and the concept of energy security will be discussed in terms of different purpose typologies of foreign policy. To clarify the subject, the widening scope of the security concept in the post-Cold War period will first be addressed; then, the extent to which environmental problems and environmental scarcity (or natural resource scarcity) have become a security problem will be explained. From this perspective, studies regarding the analysis of the subject will be evaluated, and the definition and components of energy security (the most important aspects of the relationship between natural resources and security) will be analysed. The question of what kinds of foreign policy outputs might such components create will also be addressed.

[1] PhD, Turkey Representative, Credit Europe Bank NV

## 1. The Widening Scope of the National and International Security Definitions in post-Cold War Period

At the beginning of the 1990s, shortly after the end of the Cold War, it was widely believed that the world would experience a new era of political stability, where long-term peace would prevail and conflicts would cease to exist. Francis Fukuyama's "The End of History and the Last Man" underlined that the global victory and expansion of capitalism and liberal democracy would create global peace and welfare. When we look back, it can be claimed that these evaluations stemmed, to a large extent, from the mentality and security perception of the Cold War period.

The bipolar system which had been the determining characteristic of the Cold War era constituted the main axis of this understanding of global security; thus, national security definitions were mostly made according to the position of an actor in the bipolar system and, to a large extent, within the framework of military parameters. In his evaluation of this characteristic of the Cold War, David Baldwin (1996: 119) emphasised that it not only militarised national security policies, but also militarised theoretical studies on national security; Baldwin stressed that military-based threats constituted the main axis of the national security studies of the period. Therefore, due to this security understanding and the termination of the Cold War, the aforementioned optimism began to be expressed. However, the developments encountered with the outbreak of the Gulf Crisis in 1991, and the global developments in the aftermath of that crisis, proved that the expectations were not entirely accurate, and that international security was about to face new threats and obstacles.

Many scholars have argued that these new threats cannot be evaluated within the narrow security perceptions of the Cold War period, that new security concepts and national security definitions are needed, and that it is necessary to widen the scope of security studies to emphasise focal points other than military areas.[2] In the post-Cold War period, both international security and national security definitions and scopes were addressed within a considerably wider perspective. Starting from this new

[2] For opinions in this respect, see Baldwin, Ibid, p. 123, Barry Buzan, Peoples, States and Fear: An Agenda for International Security Studies in the Post Cold War Era, London, Bibbles Ltd., and Guilford and King's Lynn, 1991, pp. 3-6, Pınar Bilgin, "Individual and Societal Dimensions of Security", International Studies Review, Vol. V, 2003, pp. 203-222, Stephen M Walt, "The Renaissance of Security Studies", International Studies Quarterly, Vol. XXXV, 1991, p. 213, Edward A Kolodziej, "Renaissance in Security Studies", International Studies Quarterly, Vol. XXXVI, December 1992, pp. 436-437.

perspective, the ground of international politics has expanded; in addition to high politics (which includes general military issues), many other problems, also referred to as low politics, have begun to be analysed as important subjects, in terms of both national and international security.

In addition to the termination of the Cold War, the driving factors and problems that changed both the national and international security understandings and scopes that are commonly expressed in many sources can be summarised as follows. It can be claimed that economic factors, above all, affected the changing scope of security perceptions. Thus, due to economic developments that were also important (but concealed) within the Cold War period, such as global economic issues, the development of countries' interdependence in terms of economic parameters, and the globalization of financial markets subject to technological developments, an economic difficulty in one country can lead to the deterioration of economic and political stability in another. The potential destabilising effect of economic variables place them within in the scope of national and international security issues. As a matter of fact, starting from the mid-2008, the economic issues arising first in the United States and then in the European Union, affected almost every country, albeit on different scales, and the economic and political consequences of this situation escalated.

The strengthening of ethnic identities suppressed during the Cold War, and the parallel increase in separatist conflicts in many nation-states, have created a widening effect on the scope of the security dimension. These developments have revived discussions concerning the phenomenon of the nation-state, and led to national security being reassessed from a different perspective. On the other hand, the fact that terrorism has gained an international character during this period reveals the need to redefine international security.

In the period following the Cold War, environmental issues constitute another set of problems, which have been expressed together with the abovementioned problems, and which have to a large extent broadened the meaning and scope of security. As well as being a set of problems in their own right, another significant characteristic of environmental issues is that they complicate the problems noted in the previous paragraphs by affecting all societies, thereby influencing the definition and scope of security.[3] In other words, environmental issues have the potential to constitute a security problem in their own right.

---

[3] For this evaluation, see Edward Page, 2000. "Theorizing the Link between Environmental Change and Security", Reciel, Vol. IX, No: 1, 2000, p. 34.

Because these issues affect countries' security perceptions and definitions as a whole, the attempts to re-define security have increased. Accordingly, security has become a comprehensive and complicated phenomenon. Hence, as Barry Buzan (1991: 19) notes, the concept of security not only has a military meaning, but also has economic, social and environmental dimensions. Concepts such as economic security, environmental security, identity security and social security, together with the military dimension of security, today constitute different forms of national security (as also emphasised by David Baldwin).

## 2. Environmental Problems, Natural Resources and the Relationship between Natural Resources and Security

### A. Environmental Scarcity and Natural Resources

In the studies conducted in the 1990s and 2000s, environmental problems, which have also affected global economic, social and political problems and whose effect increases day by day, are addressed according to two principal dimensions. The first of these dimensions is the destruction of ther environment due to the growth in world population, economic development and industrialization, and their negative effects on the ecosystem;[4] the second is the "environmental scarcity" dimension, which stems from the above-mentioned negative effects, which lead to deficiencies in the quality and quantity of natural resources.

It is claimed that environmental scarcity occurs in any part of the world due to three main factors: (a) the decrease in the quality and quantity of a natural resource caused by human factors; (b) the increase in the population; (c) the unequal distribution of resources (Izci, 1997: 413). In other words, the decrease in the quality and quantity of natural resources and environmental scarcity arise from the human factor, which is itself based on the growth of population and economic development. The unequal distribution of natural resources causes scarcity to be perceived as

[4] These effects are listed as follows in the relevant sources: a) climate change arising from greenhouse effect, b) decrease in and destruction of productive agricultural areas, c) extinction of fish species, d) stratospheric ozone layer depletion, e) decrease in and destruction of forests, f) decrease in and pollution of fresh water resources. Rana İzci, "Uluslararası Güvenlik ve Çevre", Uluslararası Politikada Yeni Alanlar Yeni Bakışlar, Ed. Faruk Sönmezoğlu, İstanbul, Der Yayınları, 1997, p. 412.

a problem, and causes this scarcity to become a factor in competition between countries; natural resources thus become a significant factor in the foreign policies of most countries. The scarcity of natural resources may thus become a national security problem.

In historical terms, natural resources constituted one of the main causes of war in Antiquity, in the Middle Ages, in the colonial era, and in the post-industrial revolution period.[5] Different natural resources came to prominence in the wars in each era. However, it can be observed that environmental problems, and the scarcity of natural resources as a national security problem, have mostly been discussed within the context of international relations (and especially in the literature on security) over the last twenty years. Accordingly, as James E. Dougherty and Robert L. Pfaltzgraff (1990: 53) note, certain factors such as geography, population, and the distribution of resources in studies on international politics and their importance in international politics are becoming increasingly important. Taking a close look at the theoretical discussions on this subject, it can be claimed that the discussions revolve around certain concepts, such as security of environment, environmental security, and natural resources security. As well as concentrating on the meanings of these three principal concepts, these discussions focus on how environmental scarcity or the scarcity of natural resources has become a security problem and, according to some, a cause of war. In other words, in addition to the essence of the problem, the meaning and content of the concepts also constitutes a main area of discussion and, as Peter Gleick (1991: 17) underlines, there is "confusion on both how to describe these concepts and how to analyse the problem".

Another issue that should be addressed at this point concerns what natural resources are, and what kinds of natural resources constitute a matter for discussion. It is emphasised in the literature that for an element to be considered as a natural resource, it should meet two prerequisites: "(a) The existence of knowledge and technical ability necessary to mine an element and to render it usable, (b) The existence of demand for the materials or services produced by element" (Rees, 1985: 13-15). With such characteristics, natural resources are examined mainly under two major topics, renewable natural resources and non-renewable natural resources.

---

[5] For further explanation, see Joseph J. Room, The Once and Future Superpower, New York, William Morrow and Company Inc., 1992, p. 160, Philippe Le Billon, "The Geopolitical Economy of 'Resource Wars,'" Geopolitics, Vol. IX, No: 1, 2004, pp. 2-4.

Renewable resources are those which can renew themselves within a certain period of time: resources such as air, water, plants, foodstuff and wind power are considered as belonging in this category. On the other hand, non-renewable natural resources consist of metals, minerals and energy resources, which are considered to have a certain quantity of "proved reserves"[6] worldwide, within the framework of existing measurements and technological levels. Oil, natural gas and coal (also known as fossil fuels), are considered to be the three main non-renewable energy resources (Rees, 1985: 238-239).

Even though the quantities of non-renewable resources may increase considerably based on technological improvements, they are considered to be essentially limited. As mentioned above, while there is a decrease in the quantity of non-renewable resources due to world population growth and economic and industrial developments, renewable resources deteriorate both in terms of quantity and quality. Due to this deterioration in quality, the renewal speed of renewable resources slows down. Moreover, the problem is complicated by the fact that renewable and non-renewable resources are not distributed equally worldwide

An important consequence of this situation is that natural resources are gaining an increasingly central place in the international relations literature. Accordingly, in the next part of the study, the primary aim will be to demonstrate the relationship between natural resources and security in international relations theory; then, the relationship between natural resources and security will be positioned within a theoretical framework.

---

[6] According to accepted definitions, the concept of proved reserve is used in the context of relating natural resources with non-renewable nature which can be economically operated with known and existing technologies. Nevertheless, it is also emphasised that the concept of proved reserve can be misleading. For instance, as William Tabb also specifies, since the quotas of OPEC are established in line with the countries' reserves, in order to increase the quotas, the member countries of OPEC may alter their reserve quantities. Or a country might want to declare its reserve quantities to be higher than they are, for political reasons. Accordingly, it is claimed that, for example, the real reserves of Russia might be less than the quantity officially expressed by the Russian government. It is also stated that the companies might be presenting their reserve quantities rather than their actual reserves, in order to keep the value of their shares high. For example, the value of Shell's shares regressed with the declaration that the company's actual reserves might be less than declared previously in 2006. For further information on this subject see William K Tabb, "Resource Wars", Monthly Review, Vol. LVIII, No: 8, 2007, p. 39.

## B. The Relationship between Natural Resources and Security in International Relations Theory

In his article "Conflict, Ecology, and the Politics of Environmental Security," Simon Dalby (2002: 125-130) specifies that the discussion in the literature on the relationship between environment, natural resources and security which took place during the 1990s resulted in a reconsideration of the security concept. Having stated that one dimension of this discussion concerned the approaches and scientific methods to be used in examining this complicated relationship, Dalby (2002: 125) also points to the multitude of definitional difficulties and the inevitability of methodological conflicts. As a result of this situation, as Dalby emphasises, such theoretical problems complicate the process of developing both research questions and political advice. In other words, it can be said that Dalby expressed the difficulties related to positioning the relationship between natural resources and security in the framework of a theoretical definition of a new field, and the determination of the place of this field within international relations theory. Despite these difficulties, it is clear that establishing an explanatory framework for the purpose of presenting the relationship between natural resources and security, by making use of different theoretical approaches, would make a significant contribution to the literature.

Approaching the problem from a historical perspective, James Dougherty and Robert L. Pfaltzgraff (1990: 54) state that the first theoretical study on the subject was conducted by Thomas Robert Malthus, in 1798. Dougherty and Pfaltzgraff state that Malthus's analysis on environmental scarcity and conflict hypothesised that although population increases geometrically, resources increase arithmetically; poverty will thus be inevitable, and population growth can only be counterbalanced by wars, famine and disease.

Also, within imperialism theory, developed by Lenin and "considered to be the most important international relations theory in the communist theory" (Dougherty and Pfaltzgraff, 1990: 230), it was claimed that one of the consequences of capitalism would be an ultimate conflict between the capitalist states over remaining markets and raw material resources. Dougherty and Pfaltzgraff claim that one of the four main elements of Lenin's theory was capitalist states' pursuit of new markets and raw material resources; according to Lenin, this would inevitably create conflicts and, as a result, would eventually create imperialist states (Dougherty and Pfaltzgraff, 1990: 54). In other words, natural resources, i.e., raw material resources, and the conflicts aimed at the control of such resources, constitute one of the main axes of Lenin's imperialism theory. Accordingly, it can be claimed that the relationship between natural resources and security has an important place in Marxist literature.

When we look at the relationship between natural resources and security within the framework of international relations, it can be seen that Realist and Neorealist theories can provide an explanatory framework. Hans Morgenthau, an outstanding exponent of classical realist theory, takes power to be the main purpose of politics, placing the concept at the centre of his analysis on international politics, and his definitiion of international politics as a power struggle (Arı, 2002: 169). For realists, power is the principal factor in states achieving their goals and realising their interests. For this reason, the struggle for power generally constitutes the foundation of international relations (Arı, 2002: 167). For realists, "national security" ranks first among the foreign policy subjects of any state; within national security, political and military issues are evaluated as high politics issues (Kauppi and Viotti, 1993: 36). For realists, other issues are secondary, being subjects of low politics. Realists also specify that it is important for a state to acquire natural resources and to maintain its access to those resources in terms of the state's national power and national security (Kauppi and Viotti, 1993: 57). In his classification of the elements of national power, Morgenthau classifies natural resources as a quantitative element.[7] From this point of view, it can be claimed that within realist theory, natural resources are regarded as a component of national power and are considered in terms of how far they influence national security. In other words, as Barry Buzan states, in realist theory natural resources are addressed in terms of national security, and in terms of the extent to which they they are necessary for a state to survive in an international system, because natural resources and energy resources are necessary for national and military powers (Buzan, 1991: 126).

As is well-known, the neorealist theory criticises classical realism on the grounds that power itself does not constitute an end but a means, and consists not only of military power but also covers economic and technological capacity (Sömezoğlu, 2005: 116). Accordingly, the neo-realists argue that the distinction made by the realist between the affairs of high politics (which revolve around military issues) and low politics

[7] Hans Morgenthau groups the elements of national power under two groups, quantitative and qualitative elements. Geography, natural resources, industrial capacity, degree of military preparedness and population are quantitative elements, while national morale, national character, the quality of diplomacy and the quality of government are considered as qualitative elements. For details see Arı, op. cit., p. 191.

(which involve other issues, of secondary importance) is meaningless (Arı, 2002: 207). One of the outstanding examples used by the neo-realists to back up this criticism concerns how the 1973 Oil Crisis, essentially an economic matter, became an issue that constituted the main agenda of international politics (Arı, 2002: 207). In other words, a natural resource could constitute a security problem and, in this sense, was able to become an issue in high politics.

Barry Buzan improved the definition of national security, which was restricted to its military security meaning by the realists, by deepening it. As mentioned above, Buzan (1991: 19) enhanced the concept of security associated only with military power within traditional realist thought, by specifying that the concept of national security does not only have a military dimension, but also has economic, societal, and environmental security pillars. In other words, the national security of a state cannot be ensured only through military power: other elements are involved. Therefore, economic security requires that a state has access to the resources necessary for its economy, and that the channels for supplying these resources are not threatened (Buzan, 1991: 242).

In the realist approach, natural resource security is perceived as a national security problem solely in terms of military power. On the other hand, for neo-realists and within the national security description (the scope of which was widened by Barry Buzan), natural resources "not only exist on the military security axis in order to fortify the military capacity, but they also fall within the economic dimension of national security due to the vital role they have for ensuring economic security" (Oktay and Çamkıran, 2006: 158).

Nevertheless, even though both these theoretical approaches provide a partial framework for showing how natural resources can create conflict, they fail to prevent conflict and, since they focus on existing structures, they remain incapable of explaining the "change" in this issue (Peters, 2004: 190). Another point that needs to be underlined is that both these approaches consider the state to be the main actor in international relations, and they therefore tend to ignore the other international actors that might play an active role in the relationship between natural resources and security. In this sense, it can be claimed that both theoretical approaches are insufficient.

Certain theoretical approaches that consider cooperation to be at the centre of international relations, and which also include many international actors in their framework analyses, present important explanatory frameworks for the purposes of defining the relationship between natural resources and security. When considered from this point of view, it can be proposed that liberal approaches, especially the pluralist perspective, can ensure a significant evolution in the study of this subject.[8]

In the field of international relations, the pluralist perspective has been described by different authors as liberalism, trans-nationalism, a world-society approach, and functionalism. The most important characteristic of the pluralist perspective worth emphasising for present purposes is that it opposes state-centric analyses that consider the subject of international relations solely via the examination of foreign policy and the relations between states. Another distinctive characteristic of the pluralist perspective, shared by almost all its proponents, is the emphasis it places on the "interdependence" arising from the fact that problems other than international security gain importance and that the foundations of the international system are changing.

Viotti and Kauppi group the main characteristics of pluralist theories (that were substantially influenced by the changes in the international system that began in the 1970s) under four main headings.[9] Above all, those who advocate the pluralist perspective distinguish themselves from the approach of realist theories that consider the state as the main actor in international relations by also accepting the existence of non-state actors. In other words, pluralist theories adopt a multi-actor understanding of international relations instead of a state-centric approach, and they include national and international pressure groups, international organizations, multi-national corporations, and even individuals in this multi-actor structure.

In the second place, pluralists do not see the state as a monolithic structure, and they claim that the state itself has sub-units and sub-organizations; for pluralists, the state is itself a pluralistic structure. Therefore, a range of actors constitute the state, such as the bureaucracy, various interest groups, public opinion, and leaders: all these are included within the pluralist perspective. Thirdly, pluralists oppose the realists who consider the state to be a rational actor; instead, they emphasise that decisions are taken as a result of a process of competition and bargaining between various actors, and in this sense they approach rationality with

---

[8] For pluralist approaches, see Kauppi, Viotti, op.cit., pp. 228-229., Arı., op.cit., s. 343-445, Helen Milner, "International Theories of Cooperation Among Nations", World Politics, Vol. XLIV, April 1992. pp. 466-496, Sönmezoğlu, op.cit., pp. 125-129, Martin Hollis, Steve Smith, Explaining and Understanding International Relations, New York, Oxford University Press, 1990, pp. 32-36, Robert O. Keohane, Joseph S. Nye Jr, "Power and Interdependence in the Information Age", Foreign Affairs, Vol. LXXVII, No: 5, September/October 1998, pp. 81-94.

[9] For all these characteristics, see Kauppi, Viotti, op. cit., pp. 228-229.

suspicion. Thus, according to a pluralist approach, decisions relating to foreign politics cannot always or solely be explained with reference to rationality.

Finally, according to pluralists, the agenda of international politics does not consist only of security matters, as realists advocate: even though security issues are of importance, other topics such as economic relations, commercial problems, rapid population growth, health, environmental issues and energy might constitute the main agenda of international relations; accordingly, the agenda of international relations diversifies. That is to say, in opposition to the realists (who consider military and security matters as the main agenda of international relations), pluralists point out that economic and social matters have become at least as important. Consequently, as mentioned by Arı (2002: 347), pluralists do not accept the distinction made by the realists between high politics and low politics.

Accepting the lack of such a distinction, and with reference to realities such as improving economic relations and achieving solutions to problems through joint actions, the pluralists proposed the concept of "interdependence" in the modern international system. In other words, one of the distinctive characteristics of the pluralist perspective is the emphasis placed on the mutual dependence between states in international politics. According to pluralists, this mutual dependence means that international relations cease to be the zero-sum game that the realists take them to be. Instead, they have a structure where actors incline towards communication and cooperation because of the additional costs which might stem from not cooperating.

The concept of "complex interdependence" depends upon the fact that for interdependence to exist between two parties, there needs to be a cost factor in the relationship or interaction (Arı, 2002: 422). That is to say, if there are no negative reflections of the interaction on the parties, it is not possible to mention interdependence. Robert Keohane and Joseph Nye categorise the conditions that create interdependence under three main headings. These are the multitude of inter-state and inter-society communication channels, the disappearance of the order of priority relating to the agenda of international issues, and, in parallel, the gradual decrease in the importance and utilization of military power.[10] When these three main points are observed thoroughly, it can be seen that the relationships between states and societies have moved far beyond formal

---

[10] For detailed explanation on these three points, see Robert Keohane, Joseph S. Nye, Power and Interdependence: World Politics in Transition, Boston, Little, Brown Company, 1977, pp. 24-29.

relations. The scope and extent of these relationships have changed with the development of communication technologies,[11] the development of organizations in which governments are not represented, and the development of multi-national corporations. Further, this multitude of actors and communication channels approximate states' policies. Secondly, according to Keohane and Nye, it cannot be claimed that military issues do not rank in first place on foreign policy agendas, in comparison with other issues. Many problems (energy problems, environment, the problem of population, and the utilization of space and open sea) have ceased to be the subjects of internal affairs, and have now become international affairs. In addition, in terms of solutions to such problems, policies that necessitate joint action will lead states towards cooperation.

According to Keohane and Nye (1977: 27-28), as a result of these first two points, military power is no longer considered an appropriate means of addressing many important problems, such as economic and ecological issues. The main factor this highlights is that the possibility of resorting to military power between states that have a complex relationship of interdependence has diminished. If this proposition is read backwards, military power is used in relationships which do not include interdependence. Accordingly, the degree of interdependence gains importance. Depending on the degree of interdependence, the option of resorting to military power might either increase or decrease. Keohane and Nye (1977: 29) also analyse the possibility of resorting to power in different circumstances. According to this analysis, if a problem does not affect vital interests deeply, the use of power is not considered. However, in a different situation, resorting to power might have a determining role to play: in such cases, it might be necessary to refer to the realist position.

The strengths and weaknesses of each approach in explaining the relationship between natural resources and security can be revealed by analysing the basic propositions of the realist and neo-realist approaches, together with those of the pluralist approaches as a whole. Above all, as will be examined in detail below, analysing the problem of energy resources (which constitute, in a way, the most important part of natural resources) and energy security solely through a perspective of conflict, as the realists and neo-realists do, might result in erroneous conclusions. As will be further explained, energy security is not only a concept valid for net energy importing countries, but also for countries that export energy

[11] For complex interdependence created by communication technologies, see Keohane, Nye Jr, op. cit., pp. 83-84.

resources. Further, the net energy exporting countries need other countries to whom they can sell their energy resources regularly at the best possible price, and energy exporting countries need investments and know-how in order to mine and operate their energy reserves (though such investment capability and know-how is possessed to a large extent by net energy importing countries). These net energy exporting countries might also need to improve their relations with transit countries, where important transmission lines (e.g., pipelines) of energy resources pass. On the other hand, for net energy importing countries, both a regular supply of energy resources at reasonable prices and the transit countries are of considerable importance. In other words, when natural resources (and, to a large extent, energy resources) are in question, it can be claimed that there exists a complex interdependence of the kind identified Keohane and Nye. For these reasons, it is clear that in addition to the potential to create conflict, natural resources and to a certain extent the energy resources (the latter constituting a significant part of the former) have the potential to create cooperation. It is also clear that owing to this potential, when analysing the relationship between natural resources and security, the strength of explanation of the realist and/or neo-realist perspectives decreases, whilst the strength of explanation of the pluralist approaches increases.

Besides, the realist and neorealist approaches, which take the state to be the principal actor, and therefore ignore other actors, render these approaches insufficient in terms of their analyses of the relationship between natural resources and security. Today, in addition to governmental institutions, many multi-national corporations play an active role in the mining and operation of energy resources, and these can be efficient in influencing and determining policies, even governmental policies. For instance, both historically and in the present day, the dominant position of many multi-national corporations constitutes a significant part of the overall evaluation and analysis relating to oil. This also applies to the production and distribution of many metals and minerals. As a matter of fact, in many examples of natural resources-based conflicts or cooperation (which at first sight seem to involve inter-state relations), it can be observed that multi-nationals have an active role. Therefore, it is clear that a theoretical approach which does not take such actors into consideration is insufficient in terms of explaining the relationship between natural resources and security. Consequently, the fact that the pluralist approach includes these kinds of actors within the scope of its analysis represents a notable contribution to the explanation of their role. However, the existence of such factors might not always create cooperation, and sometimes these actors might come to the fore as the source of conflicts.

In such circumstances, the proposition of cooperation that the pluralist approaches are founded upon might be rendered invalid.

Another point which strengthens the explanatory power of the realist position, and hence weakens the pluralist approach, is the possibility of utilising power according to whether or not the problem falls under the category of vital interests, as Keohane and Nye emphasise. It is clear that the importance of natural resources, especially metals, minerals and non-renewable energy resources, increases day by day. The increase in their importance might place this issue in a position to be evaluated via the category of vital interests. The extent to which interdependence would create cooperation in such a case remains a major question. Therefore, as Keohane and Nye also accept, it can be anticipated that the validity of the realist position will increase.

It can be concluded that the pluralist approaches, like the realist and neo-realist approaches, have certain strengths and weaknesses when it comes to explaining the relationship between natural resources and security. Therefore, as discussed by Simon Dalby, it does not in truth seem possible to address the relationship between natural resources and security within a single theoretical framework. Hence, as a result of the insufficiency of both perspectives in explaining the relationship between natural resources and security, studies should not be conducted only within the field of international relations, and studies should be concerned solely with the problem of placing the research within a theoretical framework.

## C. Studies Devoted to Explaining the Relationship between Natural Resources and Security

The studies devoted to explaining the relationship between natural resources and security have mostly been conducted in the last twenty years. Many sources state that the first and most efficient international study in the relationship between "environment" and "security" (which emphasises that environmental scarcity and natural resource scarcity might create security problems, and might even cause military conflicts) is the 1987 Final Report of United Nations World Commission on Environment and Development (the Brundtland Commission).[12] This report emphasised that environmental stress can be the cause of political tension, and specified that conflicts relating to the control of raw materials, energy

[12] For further information on this subject, see Gregory D. Foster, "Environmental Security: The Search for Strategic Legitimacy", Armed Forces & Society, Vol. XXVII, No: 3, Spring 2001, pp. 373-395.

resources, river basins, maritime areas, and other basic natural resources would increase as these resources diminish. The main factors that create the relationship between natural resources and security and that cause conflict are listed as follows: the growth of population, improper economic development and industrialization movements, poverty, and the unequal distribution of natural resources (The World Commission on Environment and Development, 1987). Many authors stress that the most comprehensive theoretical study devoted to revealing the relationship between environment and security (or natural resources and security) was conducted by Thomas Homer-Dixon.[13] In his 1999 study entitled "Environment, Scarcity and Violence,"[14] Dixon first deals with the concept of "scarcity" and states that the concept depends on three inter-related factors, which he defines as follows: supply-induced scarcity, demand-induced scarcity, and structural scarcity (Dixon, 1999: 48).

Supply-induced scarcity occurs in cases such as the deterioration of the qualities of renewable natural resources or drought; demand-induced scarcity is caused by population growth (assuming that there is an average consumption per capita); structural scarcity is the deliberate result of the action of elites (Dixon, 1999: 73). When supply- or demand-induced scarcities occur, elites mobilise and try to reorganise the distribution of resources for their own benefit. Depending on the reorganization of resource distribution, the supply- and demand-induced scarcities increase (Dixon, 1999: 73). In order to support his opinions with empirical findings, Homer-Dixon examined many case studies and conflicts, and emphasised as the main finding of his study that "significant proof is available" indicating that environmental scarcity causes violence and conflicts (Dixon, 1999: 25). The phenomenon of violence or conflict addressed in Homer-Dixon's study is, however, of a sub-national character rather than an international character. Given the study's content, it can be claimed that the international dimension of the relationship between natural resources and security was ignored.

Nevertheless, Dalby specifies that the structural scarcity in Homer-Dixon's research should be emphasised, because violence is not simply a consequence of scarcity. Rather, it stems from the elites' political, economic, and legal manipulations (Dalby, 2002: 126). In other words, all three kinds of scarcity trigger one another; however, the route to violence

[13] The authors who share this opinion are Foster, op. cit., pp. 377-390., Page, op. cit., p. 35, Simon Dalby, "Jousting with Malthus' Ghost: Environment and Conflict After the Cold War", Geopolitics, Vol. V, No: 1, 2000, pp. 165-175.

[14] For the complete study, see Thomas F. Homer Dixon, Environment, Scarcity and Violence, Princeton, Princeton University Press, 1999.

is paved by efforts to re-distribute resources, resistance to this re-distribution (structural scarcity), and the increase in the supply- or demand-induced scarcity as a result of such re-distribution. It can also be contemplated that supply- and demand-induced scarcities can be applied to the international dimension, and that the international counterpart of the elites emphasised in structural scarcity are the global elites (or the global great powers). It can therefore be said that Homer Dixon's analysis, with its international dimension, can contribute to the relationship between natural resources and security. Especially when energy resources are in question, it is clear that the control of such resources constitutes an important axis of conflict for the global powers. Accordingly, it can be envisaged that the structural scarcity approach can make a significant contribution as a result of its international dimension.

The proposition that natural resources might create security problems (and might even be the cause of wars) was also examined, although without being placed within a theoretical framework in an international dimension. In 1986, Arthur Westing (1986: 204-210) presented a list that included 12 wars and conflicts of the twentieth century, and claimed that all of these had been triggered by struggles over renewable or non-renewable resources. Westing's conclusions underlined that "the global inadequacy and deterioration of natural resources with renewable and non-renewable characteristics, together with their unequal distribution may cause unexpected alliances; national conflicts and of course wars" (Westing, 1986: 1).

It has been stated that the most comprehensive study in the international relations literature on how existing natural resources problems in the world may constitute a security problem was conducted by Michael Klare (Nevins, 2004: 257). Klare[15] (2002: 14) claims that international relations and international conflicts have gained a new

[15] Within the scope of this study, the following research by Michael Klare has been analysed: Michael T. Klare, Resource Wars: The New Landscape of Global Conflict, New York, Henry Holt, 2002, Michael Klare, Blood and Oil-The Dangers and Consequences of America's Growing Petroleum Dependency, London, Penguin Books Ltd., 2004, Michael Klare, "The New Geopolitics", Monthly Review, Vol. LV, No: 3, 2003, pp. 51-57, Michael T. Klare, Rising Powers Shrinking Planet-The New Geopolitics of Energy, New York, Henry Holt and Company, 2008., Michael Klare, "The Geopolitics of Natural Gas", Nation, Vol. CCLXXXII, No: 3, 2006, pp. 18-23, Michael Klare, Daniel Volman, "The African Oil Rush and US National Security", Third World Quarterly, Vol. XXVII, No: 4, 2006, pp. 609-628, Michael. T. Klare, "The Futile Pursuit of "Energy Security" by Military Force", Journal of World Affair, Vol. XIII, No: 2, 2007, pp. 139-153.

dimension in the period following the Cold War, and that international relations focus increasingly on keeping natural resources under control or on obtaining them. Klare says that in addition to the termination of the Cold War, and the fading into the background of ideological conflicts due to many factors, natural resources will gain in importance. The author lists these factors as: the growth of population and consumption in many regions of the world, the fast pace of industrialization and the natural resources this demands, the monetary value of some special natural resources, and the existence of inter-state struggles and conflicts over critical natural resources, particularly those located at frontier regions and offshore areas (2002: 14-15).

Klare (2002: 29) points to oil and natural gas as the natural resources that have the greatest potential to create war in the twenty-first century. Stating that oil and natural gas will continue to have the same effects thay had in the 20th century, Klare specifies that the Persian Gulf (the main region of oil supply and distribution) may become a significant area of conflict, along with other areas including the Caspian Sea Basin and the South China Sea. The common characteristics of these three areas that might create conflicts are that their borders, where oil and natural gas resources are available, are problematic (overland and at sea), the local political regimes are authoritarian, local resources are distributed unequally, small local elites manage these resources, and the long history of regional-ethnical hostilities makes it likely that external interventions will fuel these fires (Klare: 2002: 35-47).

Klare (2003: 52) specifies "water" as another natural resource that might cause conflicts in addition to oil and natural gas. He stresses that decreasing water resources, as a consequence of fast population growth and domestic political developments as well as global warming and the existence of problematic cross-border water resources, will increase conflict, as with oil and natural gas. Klare sees the Middle East and south-west Asia as the probable conflict areas in terms of water. The author also states that in many regions of the world, timber and precious minerals may constitute causes of wars. Southeast Asia and, on the American continent, Brazil, Mexico and Colombia, as well many central African countries, are highlighted as the principal locations of such conflicts (Klare, 2003: 55).

Within the framework of his studies, Klare (2003: 56-57) created a global map showing the possible areas of conflict based on natural resources, and claimed that this indicated the principal fault lines in international relations. He says that many problems are interconnected in the conflict areas on this map, and that this situation renders the problems and conflicts related to natural resources more severe. As the solution to these natural resource based conflicts, Klare (2002: 223) proposes that the map should be considered as a whole: thus, alternative energy resources and industrial implementations should be researched through global cooperation and by international institutions, and a strategy offering a more equal distribution of existing natural resources should be developed.

Compared with other studies concerning how natural resources create security problems, Klare's work has made a significant contribution to the literature. He tries to include all natural resources and adopts a global point of view. Furthermore, he emphasises seceral different aspects, including the following:

a) The influence of domestic political developments (the existence of authoritarian regimes and of political instability).
b) The existence of regional problems.
c) The existence of regions that have natural resources on which different states have claims of sovereignty (such as frontier regions, rivers, and offshore areas).
d) Utilisation of natural resources as a weapon (especially water resources and/or oil and natural gas).
e) The intervention of global powers (external factors).

Addressing the question of how natural resource scarcity might cause security problems can be regarded as an important contribution to the effort to establish a theoretical framework. On the other hand, Klare's work lacks comprehensive research into the solution to conflicts, and lacks a detailed proposal regarding this matter.

To summarise, the following factors are shared by the above-mentioned studies relating to how natural resources may create national and international security problems. In other words, the factors below come to prominence when establishing a theoretical framework related to the relationship between natural resources and security. These factors could be used as an explanatory framework for analyses concerning natural resources constitute security problems.

**a) Objective Factors:**
1) Scarcity factor: Deterioration of the quantity and quality of renewable and non-renewable natural resources worldwide as a result of fast population growth, economic development, and industrialization.
2) Distribution factor: The unequal distribution of natural resources worldwide.
**b) Subjective Factors:**

1) The possibility of utilising natural resources, in a sense, as a weapon in global power struggles, and the intervention of global powers in the distribution and utilization of natural resources.
2) The potential for natural resource problems to affect regional problems, or vice versa.
3) The effect of domestic political structures and regimes on the distribution of natural resources, and potential for this situation to constitute a problem.
4) The existence of natural resources regions upon which different countries claim sovereignty.

Any one of these factors might come to the fore in different problem situations relating to natural resources and security. The level and explanatory power of each factor might therefore differ, depending on the problem being addressed. It can also be stated that for a problem related to the scarcity of a natural resource, a new subjective factor that might be added to the objective factors could accelerate the transformation of the issue into a security problem. Any subjective factor added on the objective factors might therefore play an important role in the transformation of the issue into both a national and international security problem.

Owing to the increasing influence of these factors in natural resource scarcity problem, it can be said—and this is mentioned by David Baldwin—that different forms of natural resource security (that are also included in different forms of national security) have emerged, and that concepts like food security, water security and energy security have started to gain a place in our understanding of natural resources security. In fact, these concepts have pushed the concept of natural resources security into second place. Accordingly, the share of these security concepts has increased rapidly in the international relations and international security literature, and it can be observed that under the concept of "resource wars," new concepts such as "food wars," "water wars," and "energy wars" are now being discussed, and scenarios concerning these matters are being examined with increasing frequency. In other words, the concept of natural resource security is segmenting, and a differentiation in the literature is being created through studies concerning one or another specific natural resource. Consequently, in contrast with those studies that twenty years ago covered all natural resources, it can be predicted that the development of the literature from now on will be towards studies that deal with specific natural resources.

Hence, the same development has been valid for energy resources and the concept of energy security, which emerged within the investigation of the concept of natural resources security. Accordingly, in the next section, the development of the concept of energy security will be analysed. In order to render it functional, the concept will be explained by presenting a range of opinions related to it.

## 3. Energy Security within Natural Resources Security

### A. Definition and Components of Energy Security

It can be claimed that in both Michael Klare's studies and many other studies analysed within the scope of this paper, the most important concept discussed in terms of the relationship between natural resources and security is energy security. All of these studies give priority to energy resources, whilst listing those natural resources that might cause security problems. Each of these studies includes the concept of energy security. Some authors even emphasise energy resources as the sole resource, and energy security as the sole problem of security related to natural resource security and to natural resource wars (Peters, 2004: 187). Within the literature, it is noticeable that especially in the last ten years, the concept of energy security has been treated and analysed in isolation from the concept of natural resources security. The importance attached to energy security has increased beyond the security of natural resources. In other words, it can be claimed that the concept of energy security has to a certain extent gained a place of its own in the security literature. Thus, in this part of the study, the meaning and components of the concept will be explained.

It is claimed by many authors that energy began to be evaluated as an element of security, and that the history of energy security started during the Cold War, especially after the oil crisis of 1973-1974.[16] OPEC used oil as a dissuasive element during the crisis, and the effects created by this action proved the importance of providing energy supply to countries. In parallel, the OECD countries established the International Energy Agency in 1974, with an emphasis on the concept of energy security (Belyi, 2003: 355). Henry Kissinger, the then US Secretary of State, declared in 1975 that Washington might go to war because of oil, underlining the concept of energy security due to an oil crisis (Klare, 2002: 33).

The second phase of this process is the Carter doctrine, named after Jimmy Carter, the then President of the US, as a result of Russia's invasion of Afghanistan in 1979. Carter indicated in 1980 that the US

[16] For further information, see Nevins. op. cit., p. 257, Klare, op. cit., p. 33.

would use all necessary means to prevent the hostile movements that might limit the flow of oil through the Persian Gulf (Nevins, 2004: 257). It is thought that together with energy security, the concept of "resource wars" was used for the first time in the literature at the beginning of the 1980s in the US, within the scope of the evaluations related to the Carter Doctrine (Billon, 2004: 188).

The broad utilization of the concept of energy security in the literature started in the post-Cold War period and has since increased, especially in the last ten years. The studies devoted to revealing the meaning and the components of the concept therefore have a recent history. When the concept of "energy security" is examined closely, it can be seen that the meanings attributed to the concept vary within the relevant literature, and that different evaluations are presented according to criteria such as whether countries have energy resources, the economic development levels of the countries, their proximity to energy resources, and whether their energy resources are located on trade routes.[17] Additionally, and since "energy" is the main input for countries to continue to exist and develop, the fact that energy resources have become an increasingly significant dimension of global power struggles has tended to diversify the the proposed definitions and evaluations of the concept.

For instance, the meaning of energy security for an oil exporting and importing country differ and, for oil importing countries, different components regarding the concept might come to the fore, such as the level of economic development, its position in global power struggles, or its position as a transit country on energy routes. These cases might be considered as the elements which complicate the definition of the concept, making it difficult to address within a theoretical framework. Perhaps

[17] For various evaluations in studies devoted to explaining the concept of energy security, see Daniel Yergin, "Ensuring Energy Security", Foreign Affairs, Vol. LXXXV, No: 2, Mar-April 2006, pp. 69-82, Gareth M. Winrow, "Energy Security in the Black Sea Region: Economic Interdependence or Commercial and Political Rivalry?", Southeast European and Black Sea Studies, Vol. I I, No: 2, May 2002, pp. 129-152, Tim Lloyd Wright, "On Seeking Energy Security". Hydrocarbon Processing, September 2007, pp. 13-14, John M. Roberts, "The Black Sea and European Energy Security", Southeast European and Black Sea Studies, Vol. VI, No: 2, 2006, pp. 207-223, Necdet Pamir, "Kafkaslar ve Hazar Havzasındaki Ülkelerin Enerji Kaynaklarının Türkiye'nin Enerji Güvenliğine Etkileri", 'Türkiye'nin Çevresindeki Gelişmeler ve Türkiye'nin Güvenlik Politikalarına Etkileri' Sempozyum Tutanakları, İstanbul, Harp Akademileri Yayınları, 2006, pp. 1-76, Nader Elhefnawy, "Toward a Long Range Energy Security Policy", Parameters, 2006, pp. 101-114.

because of this characteristic, defining the concept has been avoided and many studies have focused only on its increasing importance.

Another aspect of the subject that should be emphasised, before any attempt is made to define the concept, is the question of which energy resources are included in the concept of energy security. Oil, natural gas, and coal (also called fossil resources), are the natural resources that are expressed in terms of energy security in almost all existing studies. However, apart from these resources, all the natural resources that could diversify energy production and increase energy production levels can be analysed within the scope of energy security. As will be explained below, this is because one of the most important parameters for ensuring energy security is diversifying resources. Accordingly, hydroelectric power, wind power, and geothermal power could be analysed within the scope of energy security, to the extent that they contribute to resource diversification and influence energy-based policy. On the other hand, it can be envisaged that the effectiveness of these resources in terms of energy security will be limited, because their present and future role in the energy potential of the world are also limited. However, this situation will differ from one country to another.

Even though the attempt to define energy security is avoided in many studies, many authors have made efforts to define the concept by addressing its various dimensions. If we examine these studies' definition of the concept, we find that energy security is considered above all to be the equivalent of supply security, and the existence of supply facilities of energy resources is evaluated as the sole element of energy security. Such definitions can be qualified as those in which the supply dimension of the concept is brought to the fore. Some studies add the finance factor to supply security, and emphasise the fact that energy supply at a reasonable price is also important in terms of energy security. In other studies that elaborate other dimensions, not only the energy security of the net energy importing countries is considered, but evaluations related to the energy security of the net energy exporting countries are also made. In other words, there is more than one definition of energy security, and each definition focuses upon different dimensions and components of the concept.

If these definitions are analysed as a whole, it is clear that, primarily, the opportunities of supply and its maintenance constitute a very important dimension in ensuring energy security. However, the most basic assumption in many of the definitions which consider energy security as the equivalent of supply security is that energy security is a concept valid only for net energy importing countries. On the other hand, it is known that net energy exporting countries also use the concept of energy security, and have in recent years increasingly talked about their own energy security. For this reason, definitions which claim that "energy security is the equivalent of supply security" have a limited scope, due to their emphasis on the security of supply. Considering the issue only through the supply dimension might also have a limited capacity to reveal the importance of energy security to a country, because in most definitions only the supply dimension of energy and the political dimension of the matter are emphasised. On the other hand, for energy importing countries, in addition to the political aspect, pricing and finance are also important elements. This is because for most of these countries, providing energy products is an important component of their gross national product. Price increases or decreases affect the economy of the country positively or negatively, and it is clear that such effects may have significant consequences in terms of security considerations. Recently, for instance, one of the most important dimensions of the Russia-Ukraine energy crises, also important in terms of the energy security of the EU, was the lack of agreement on prices. Following this, one of the issues most frequently discussed as regards the import of natural gas to Turkey is the cost of procurement. The cost of energy resources also constitutes a very important issue for net energy exporting countries, because a considerable part of their economies consist in the income obtained from the sale of energy products; these incomes might even be a significant factor in ensuring the legitimacy of political regimes. On the other hand, it can also be envisaged that as price levels increase, the level of economic and political influence of energy exporting countries will also increase. Correspondingly, it is clear that (for energy exporting countries) the price dimension and (for the energy importing countries) the finance dimension constitute important factors in terms of energy security.

It can thus be understood from these evaluations that energy security is a concept utilised by both net energy exporting and net energy importing countries; the concept thus includes both economic and politic components. In conjunction with this, it can also be argued that the most comprehensive study for the purpose of presenting different evaluations concerning energy security (and, accordingly, the different definitions and components of the concept) was conducted by the World Bank.[18] In this study, the World Bank proposed its own definition of energy security, emphasising, which emphasised that energy security means "ensuring

[18] For the full study, see The World Bank Group, Energy Security Issues, World Bank Working Paper Series, No: 36110, Washington D.C., December 2005.

countries can sustainably produce and use energy at reasonable cost in order to facilitate economic growth and, through this, poverty reduction, and directly improve the quality of peoples' lives by broadening access to modern energy services" (The World Bank Group, 2005: 3). However, after offering this definition, the World Bank specified that different countries possess different definitions, and established a matrix for such different definitions within the framework of the principal criteria. The two principal criteria upon which the World Bank establishes its matrix are the level of economic development and the endowment of energy resources (The World Bank Group, 2005: 3). The table formed by the World Bank is as follows:

**Table 3-10 Perspectives of Different Countries on Energy Security of the World Bank**[19]

| Endowment of Energy Resources and Level of Economic Development | Definition and Components of Energy Security, and Priorities in Energy Security |
|---|---|
| Industrialised Net Energy Importer Countries | *Avoid disruption of energy supplies<br>*Diversification of energy supply sources<br>*Security concerns for energy infrastructure<br>*Technological solutions to reduce dependence on imported supplies. |
| Major Hydrocarbon Exporting Countries | *Creation of long term markets with reasonable prices<br>*Diversification of export markets for energy resources<br>*Securing capital and financing for investment in resource development and infrastructure. |
| Large Emerging Markets with Rapidly Growing Energy Demands | *The ability to meet growing demand for energy from imported sources<br>*Diversification of energy supply sources<br>*Securing financing for investment in resource development and infrastructure.<br>*Development of technological solutions to reduce dependence on imported supplies.<br>*Meeting people's basic energy needs and creating effective demand for energy services. |
| Mid-Income Net Energy Importer Countries | *The ability to meet growing demand for energy from imported sources<br>*Securing financing for investment in resource development and infrastructure.<br>*Meeting people's basic energy needs and creating effective demand for energy services |
| | *The ability to meet growing demand for energy from imported sources |

[19] For the complete table see Energy Security Issues. Ibid, p. 4.

| Low Income Net Energy Importer Countries | *Securing financing for investment in resource development and infrastructure.<br>*Meeting people's basic energy needs and creating effective demand for energy services |
|---|---|

On examination of this table, it can be seen that there are some factors which are not clarified, due to the assumptions upon which the table is based. However, these need to be included within the definition of the concept of energy security. Above all, the most important omission in the World Bank's study is the political dimension. As a result of this, the place of energy security within global power struggles and the provision of the concept of energy security in terms of different purpose categories in foreign politics are both ignored. For example, the policies of the US and of any EU country, which are all net energy importer countries, may vary in the provision of energy security; while the US, which considers energy as an element of a global power struggle (control of energy resources, transit routes of energy and energy trade), might chose a series of policies for energy that could lead to the use of force, for an EU country the same might not be possible. In other words, in the terminology of the World Bank, different perspectives might also occur also amongst "developed net energy importer countries"; again, due to the lack of a political dimension, the case of "major hydrocarbon exporting countries" having different perspectives on energy security is also ignored. For example, it might be misleading to claim that Norway and Russia, both net exporters, would have the same definition of "energy security" and that they would apply and defend the same policies.

It can be claimed that despite this basic lack, the World Bank study fills an important gap, by introducting the principal elements of energy security. In terms of formulating a functional definition of energy security, by considering both the above-mentioned studies conducted by the World Bank and the other works related to the subject, two different definitions are possible: one for the net energy exporting countries, and the other for the net energy importers. Accordingly, it can be said that for the **net energy importing countries,** the following policies devoted to the provision of energy constitute the main elements of energy security policies:

a) From **diversified sources,** in terms of method of production and/or country of origin.
b) In an **interrupted** way (both the interruption of the source and the provision of interruption on the transit routes of the resource).

c) From **reliable source countries** in terms of political relationships.
d) With the **timing appropriate** to the energy demand.
e) With **reasonable costs** at financeable levels.

For net energy exporting countries, policies devoted to the following points constitute the main elements of energy security policies:

a) **Diversification of export markets** of energy resources.
b) Maintaining export markets at **high price levels and over the long term**.
c) In order to maintain production and export levels over the long term, investment in the necessary infrastructure.
d) Starting from the fact that energy is the main input in the development and survival of all economies, to benefit from such a power as leverage in developing other economic and political relationships, and/or in creating economic and political influence on export markets.

For net energy exporting countries, just for as net energy importing countries, the policies established in order to diversify the resources used for energy production can be evaluated as policies devoted to increasing energy security.

It can be seen that the elements constituting the above-mentioned definitions have economic and political dimensions, as well as those dimensions covered by international relations. In addition to these dimensions, another aspect that should be taken into consideration within the concept of energy security is the fact that attention is increasingly being paid to environmental concerns when defining energy security and constructing policies associated with such definitions. It is clear that despite their contribution to economies, energy production and consumption processes have adverse effects on environmental pollution and global climate change. Therefore, efforts to minimise the environmentally adverse effects of energy policies, or shaping energy policies with a view to not harming the environment, are meaningful in terms of impact of the environment on the definition of countries' energy security. Hence, it can be observed that environmental factors have significant influence on policies aimed at increasing the share of renewable energy resources in the energy consumption of some countries, or on the global increase in the utilization of natural gas. The need to include environmental concerns in the definition of energy security is specified by many authors.[20]

It can be claimed that due to all these factors that broaden the definition, it is misleading to try to explain the concept of energy security within a single international relations theory, or within a single economic theory. Therefore, as was earlier suggested when explaining the relationship between natural resources and security, different theoretical approaches might have different strengths and weaknesses for the purpose of explaining energy security. Hence, quoting from Susan Strange, Andrei Belyi (2003: 353) emphasises that the existing theoretical limits of political science and international relations theories are insufficient for the purpose of analysing the concept of energy security; instead, the concept should be analysed as a whole, and its economic and political dimensions should be included.

Accordingly, Andrei Belyi specifies that the concept of energy security can be explained via a structural approach that is specific to international political economy[21] (in Belyi's definition, according to this approach, structures define the global political and economic behaviours of states, companies, and other social and economic actors). According to Belyi, four principal structures constitute the structural power of international actors. These are security, finance, production, and knowledge. Therefore, energy constitutes the structural power of international actors in terms of the following dimensions:

a) **Production Dimension** (due to utilization of energy in all industries, transportation and dwellings).
b) **Financial Dimension** (due to the importance of the income obtained from the energy resources, or the need of countries for substantial financial power in order to buy energy resources).
c) **Security Dimension** (due to the ability of energy resources to be the source of military interventions and even wars).
d) **Knowledge Dimension** (since acquisition of energy resources is possible through a technology that is substantially knowledge-intensive) (Belyi, 2003: 353).

---

[20] For such opinions, see Örgen Uğurlu, Çevresel Güvenlik ve Türkiye'de Enerji Politikaları, İstanbul, Örgün Yayınevi, 2009, pp. 107-111, Gawdat Bahgat, "Energy Security: What Does It Mean? And How Can We Achieve It?", The Journal of Social, Political and Economic Studies, Vol. XXXIII, No: 1, Spring 2008, p. 87.

[21] The original concept utilised by Andrei Belyi is the concept of "international political economy." For its utilization, see Belyi, loc. cit., p. 353.

The concept of energy security is influenced by these four dimensions. This renders the definition of energy security difficult, limits the establishment of a theoretical framework, and renders the presentation of the concept via a multi-disciplinary approach obligatory. Owing to these difficulties, it seems that a significant gap exists in the discussions in the literature on energy security and the establishment of a theoretical framework.

Even though the elements of energy security presented above for the net energy importing and net energy exporting countries constitute a whole, any one element might come to prominence either for a net energy importer or exporter, and the energy security policies for such a country might be defined by considering that element as the main axis. Energy security policies might differ according to the different elements that come to prominence.

It can also be claimed that another reason why different energy security policies occur is because energy security policies might be formulated in terms of different typologies of purposes in foreign policy. Therefore, within the scope of this study, it is important to offer an answer to the question of what kinds of purposes energy security might be subject to.

## B. Energy Security in terms of Different Purpose Typologies in Foreign Policy

As highlighted by Mehmet Gönlübol (1993: 91), political actors in the international arena do not react solely to external actions, nor do they simply adapt to new international situations. Thus, it can be claimed that the most important political actors on the international arena act according to a purpose and/or a set of purposes. Starting from this perspective, various studies have been conducted as to what these purposes might be, and in order to define such purposes many classifications have been proposed around different criteria.[22]

Faruk Sönmezoğlu (2005: 271) specifies that the foreign policy purposes of a state might be listed under three general categories: a) purposes related to existence; b) purposes in the range of security and prestige; c) long term/ geopolitical and ideological purposes. As also underlined by Sönmezoğlu (2005: 271), it can be claimed that by definition, every nation has the fundamental aim of existence and of

---

[22] For classifications based on different criteria, see Faruk Sönmezoğlu, Uluslararası Politika ve Dış Politika Analizi, 4. bs., İstanbul, Filiz Kitabevi, İstanbul, 2005, p. 269.

maintaining its existence. Even though the meanings of "existence" and "maintaining its existence" are not very clear, as Mehmet Gönlübol (1993: 94) points out, it is possible to say that there is a general consensus that the most important value related to the maintenance of existence in foreign policy is the "defence of homeland." Today, it does not seem very plausible to consider energy security policies within this category. However, in the long term, due to the possibility of energy scarcity (which it is thought will increase in the future) and conflicts which might arise amongst states with the purposes of resolving such scarcity, it can be claimed that energy security policies might have the capacity to create consequences that threaten the existence of states. In other words, to the extent that the importance of energy increases in the long term, the issue of energy might be evaluated by some countries as a dimension of existence.

As to the wide range of purposes related to security-prestige, Faruk Sönmezoğlu (2005: 273) analyses these within the framework of three categories: security, vital interests/spheres of influence, and prestige. The purpose of security is defined as being "very closely related with the purpose of existence" and as "taking the measures which render the defence of homeland directly not necessary any more" (Sönmezoğlu, 2005: 274). As mentioned above, while explaining the broadening of the scope of security in the post-Cold War period, it can be said that to a large extent, the concept of energy security can be considered as a category of foreign policy. However, in her study devoted to presenting the concept of security at different levels (such as national security, international security and global security), Helga Haftendorn (1991: 3) specified that the concept of security is not clear. The existence of different considerations and different definitions related to energy security (as discussed above) when trying to explain the concept of energy security justifies Haftendorn's argument.

The concept of vital interests, which is intricately related to the concept of security, is classically defined as "the purposes for which one is ready to fight" (Sönmezoğlu, 2005: 278). It is clear that energy security is not a vital interest for many net energy importing countries. However, as the level of dependence on energy increases, and procurement of resources becomes more difficult, it can be asserted that in theoretical terms, this dimension of security could evolve towards the category of a vital interest. From this perspective, taking the levels of energy utilization, procurement and dependence as a basis, it is clear that in the long run, the number of countries which position energy security in the category of vital interests will also increase. In other words, in the long term, energy security possesses the potential to be transformed into a purpose "worth fighting for."

In the category of creating spheres of influence and maintaining their existence by creating spheres of influence, it is clear that this foreign policy purpose might be valid for net energy exporting countries. However, this might be evaluated alongside a country's other elements of power. In this context, it becomes apparent that the spheres of influence and the purposes of creating spheres of purposes of, for instance, Russia or Saudi Arabia (which are both net energy exporters), would be different. On the other hand, as energy resources diminish, it is to be expected that energy policies creating spheres of influence and/or maintaining existing influence will be increasingly used by the net energy exporting countries. In other words, the capacity of energy resources to create influence will increase, and such influence will be used more and more by those countries that possess energy resources.

It is specified that there might be two main types of long term purpose in the foreign policy of states: geopolitical purposes and ideological purposes (Sönmezoğlu, 2005: 281). When considered from the perspective of energy policies (especially policies related to possessing energy resources), controlling transit routes and trade in such resources can be regarded within the category of long term geopolitical purposes. Starting from this perspective, the energy security policy of the US might be considered as falling within the category of long term geopolitical purposes. Moreover, Russia's energy policies also have the characteristics fitting this category, since they can be considered as a means of achieving geopolitical ends.

It can be seen that energy security and policies devoted to the provision of energy security can be analysed within several categories of foreign policy purposes. It is also clear that, depending on the increasing importance of the matter, the place of energy security has been ascending in the ranking made by both net energy importing and net energy exporting countries, relative to their foreign policy purposes. In other words, as the scarcity of energy resources escalates, and as the levels of dependence on external energy sources increase, the importance attached to the purposes related to energy security will also increase. Given such features, the subject of energy security constitutes a significant subject for discussion in international relations, and can be considered an integral and important part of national security.

Another issue that should be addressed is that although different definitions are valid for net energy exporting and net energy importing countries, such countries develop different policies within the framework of these definitions. The place of energy in the ranking, related to foreign policy purposes, ascends as a consequence of the increasing importance of the matter. Accordingly, the matter becomes a part of the national security policy. Thus, it might be misleading to claim that energy policies have a structure which could directly create conflicts. For, as specified by Gawdat Bahgat (2006: 89), interdependence is today an inevitable reality within the global energy system.

Due to this interdependence, and even though big energy producer countries 9or those possessing energy resources) desire to utilise resources in such a way as to maximise their returns economically and politically, the implemented policies emerge as a result of a complicated bargaining process. The factors that create this bargaining process are multi-dimensional. Above all, net energy exporting countries have the basic goal of maintaining the markets in which they regularly sell their energy products. Accordingly, the goal of maintaining these regular markets creates the effect of bringing producers and consumers together. Additionally, especially in the transmission of oil and natural gas through pipelines, transit countries become a part of the game, and can be influential in the bargaining process. Another factor that renders the subject more complicated is the existence of a structure in which private companies play a part. In the production and transmission of primary energy resources such as oil, natural gas, and coal, the policies of such companies might cause countries to make different concessions. Or, on the contrary, these companies might have to make concessions in terms of their profitability. The main issue highlighted by these matters is that the various definitions of energy security and different policies developed according to these definitions cannot be considered as a zero-sum game.

Consequently, evaluating the concept of energy security within the framework of the above explanations and discussions, it seems possible to establish the following findings:

* The development of the concept began as part of the concept of natural resources. However, especially in the 2000s; it gained a place beyond natural resources security, and was explored in a much broader way in the literature.

* Even though there have been attempts to define the concept in the literature, consistency has not been achieved, and many definitions which address different dimensions and components of energy security are available.

* The existence of different dimensions of the concept of energy security, which are included in the fields of economy, politics, environment and international relations, renders the definition of the concept difficult. At the same time, this situation shows the existing theoretical limits of economics, political science and theories of international relations in terms of analysing the concept of energy security. This, in turn, constrains the establishment of a theoretical framework and gives rise to the need to define the concept via a multi-disciplinary approach.

* The energy security policies of countries may vary depending on the emphasis placed upon different dimensions of the concept, and this causes the meaning given to energy security by each country to be different. It is possible to analyse energy security policies in terms of different categories in foreign policy purpose, depending on the importance of the matter. The place of energy security has been ascending in the ranking made by countries, related to their foreign policy purposes.

## References

"Individual and Societal Dimensions of Security", International Studies Review, Vol.V, 2003, pp. 203-222.

Arı, Tayyar, Uluslararası İlişkiler Teorileri, Cağaloğlu/İstanbul, Basım Yayım Dağıtım Ltd. Şti, 2002.

Bahgat, Gawdat, "Central Asia and Energy Security", Asian Affairs, Vol. XXXVII, No: 1, March 2006, pp. 1-16.

—. "Central Asia and Energy Security", Asian Affairs, Vol. XXXVII, No: 1, March 2006, pp. 1-16.

—. "Energy Security: The Caspian Sea", Minerals&Energy, Vol. XX, No: 2, 2005, pp. 3-15.

—. "Energy Security: What Does It Mean? And How Can We Achieve It?", The Journal of Social, Political and Economic Studies, Vol. XXXIII, No: 1, Spring 2008, pp. 85-98.

—. "Europe's Energy Security: Challenges and Opportunities", International Affairs, Vol.LXXXII, No: 5, October 2006, pp. 961-975.

—. "Pipeline Diplomacy: The Geopolitics of the Caspian Sea Region", International Studies Perspectives, Vol. III, 2002, pp. 310-327.

—. "Regional Report-The Caspian Sea: Potentials and Prospects", Governance, Vol. XVII, No: 1, January 2004, pp. 115-126.

—. "Russia's Oil Potential: Prospects and Implications", OPEC Review: Energy Economics&Related Issues, Vol. XXVIII, Issue 2, June 2004, pp. 133-147.

—. "Terrorism and Energy", World Affairs, Vol.CLXVII, No: 2, Fall 2004, pp. 51-58.

—. "The Geopolitics of Natural Gas in Asia", OPEC Review: Energy Economics&Related Issues, Vol. XXV, No: 3, September 2001, pp. 273-291.

—. "United States Energy Security", The Journal of Social, Political and Economic Studies, Vol. XXVI, No. 3, 2001, pp. 515-542.

Baldwin, David A., "Security Studies and the End of the Cold War", World Politics, Vol. XXXXVIII, No: 1, 1996, pp. 117-141.

—. "The Concept Of Security", Review of International Studies, Vol. XXIII, 1997, pp. 5-26.

Barcella, Mary L., "Natural Gas in the Twenty-First Century", Business Economics, Vol. XXXI, No: 4, 1996, pp. 19-24.

Barnes, Joe, Amy Jaffe, Edward Morse L., "The New Geopolitics of Oil", National Interest, Winter 2003-2004, pp. 7-15.

Barnett, Jon, Uluslararası İlişkiler Teorileri, Cağaloğlu/İstanbul, Basım Yayım Dağıtım Ltd. Şti, 2002.

Barton, Barry, Catherine Redgwell, Anita Ronne, Donald N. Zillman, Meaning of Environmental Security: Ecological Politics and Policy in the New Security Era, London and Newyork, The Zed Boks, 2001.

Belyi, Andrei V., "New Dimensions of Energy Security of the Enlarging EU and Their Impact on Relations with Russia", European Integration, Vol. XXV, No: 4, December 2003, pp. 351-369.

Bilgin, Pınar, "A Return to 'Civilizational Geopolitics' in the Mediterranean? Changing Geopolitical Images of the European Union and Turkey in the Post-Cold War Era", Geopolitics, Vol. IX, No: 2, 2004, pp. 269-291.

—. "Turkey's Changing Security Discourses: The Challenge of Globalization", European Journal of Political Research, Vol. XXXXIV, 2005, pp. 175-201.

Billon, Philippe Le, "The Geopolitical Economy of 'Resource Wars'', Geopolitics, Vol. IX, No: 1, 2004, pp. 1-28.

Billon, Philippe Le, Fouad El Khatib, "From Free Oil to Freedom Oil: Terrorism, War and US Geopolitics in the Persian Gulf', Geopolitics, Vol. IX, No: 1, March 2004. pp. 109-137.

Brown, Michael, Brave New World: Security Challenges in the 21th Century, Washington, Georgetown University Presss, 2003.

—. New Global Dangers: Changing Dimensions of International Security, The MIT Press, 2004.

Browne, John, Vivienne Cox, Proceedings of Energy Security: Responding To The Challenge, Washington, The Brookings Institute D.C., 2005.

Buzan, Barry, Peoples, States and Fear: An Agenda for International Security Studies in the Post Cold War Era, London, Bibbles Ltd., and Guilford and King's Lynn, 1991.

Caldwell, Dan, Robert E Williams, Seeking Security in an Insecure World, Rowman&Littlefield, 2006.

Dalby, Simon, "Conflict, Ecology, and the Politics of Environmental Security". Global Environmental Politics, Vol. II, No: 4, 2002, pp. 125-130.

Dalby, Simon, "Exorcising Malthus's Ghost: Resources and Security in Global Politics", Geopolitics, Vol. IX, No: 1, 2004, pp. 242-254.

—. "Jousting with Malthus' Ghost: Environment and Conflict After the Cold War", Geopolitics, Vol. V, No: 1, 2000, pp. 165-175.

Debra, Johnson, "EU-Russia Energy Links: A Marriage of Convenience?", Government&Opposition, Vol. 40, Issue: 2, Spring 2005, pp. 256-277.

Dixon, Thomas F. Homer, Environment, Scarcity and Violence, Princeton, Princeton University Press, 1999.

Dougherty, James E., Robert L. Pfaltzgraff, Contending Theories of International Relations, New York, Harper&Row Publishers, 1990.

Flavin, Christopher, Seth Dunn, "A New Energy Paradigm for the 21th Century", Journal of International Affairs, Vol. LIII, No: 1, 1999, pp. 167-190.

Foster, Gregory D., "Environmental Security: The Search for Strategic Legitimacy", Armed Forces&Society, Vol. XXVI, No: 3, Spring 2001, pp. 373-395.

Gleick, Peter H., "Environment and Security: The Clear Connections", The Bulletin of The Atomic Scientists, April 1991, pp. 17-21.

Goldstone, Jack A., "Scarcity, Crises and Choice", Journal of International Affairs, Vol. LIX, No: 2, Summer 2006, pp. 335-346.

Gönlübol, Mehmet, Uluslararası Politika-İlkeler, Kavramlar, Kurumlar, Ankara, Atilla Kitabevi, 1993.

Haftendorn, Helga, "The Security Puzzle: Theory Building and Discipline Building in International Security", International Studies Quarterly, Vol. XXXV, No: 1, March 1991, pp. 3-17.

Hill, Fiona, Energy Empire: Oil, Gas and Russia's Revival, London, The Foreign Policy Centre, 2004.

Hollis, Martin, Steve Smith, Explaining and Understanding International Relations, New York, Oxford University Press, 1990.

Houweling, Henk, Mehdi Parvizi Amineh, "The Crisis in IR Theory: Towards a Critical Approach", Perspectives on Global Development &Technology, Vol. III, No: 3, 2003, pp. 315-335.

Işıksal, Hüseyin, "To What Extend Complex Interdependence Theorists Challenge to Structural Realist School of International Relations?", Alternatives: Turkish Journal of International Relations, Vol. III, No: 2-3, Summer-Fall 2004, pp. 130-156.

Kalicki, Jan H., David L. Goldwyn, Energy and Security: Toward a New Foreign Policy Strategy, USA, Woodrow Wilson Center Press, 2005.

Kauppi, Mark V., Paul Viotti R., International Relations Theory: Realism, Pluralism, Globalism and Beyond, New York, Allyn and Bacon, 1993.

Kelly, Phil, "A Critique of Critical Geopolitics", Geopolitics, Vol. 11, 2006, pp. 24-53.

Keohane, Robert O., Joseph S. Nye Jr., "Power and Interdependence in the Information Age", Foreign Affairs, Vol. LXXVII, No: 5, September-October 1998, pp. 81-94.

Keohane, Robert, Joseph S. Nye, Power and Interdependence: World Politics in Transition, Boston, Little, Brown Company, 1977.

Klare, Michael T., Kaynak Savaşları: Küresel Çatışmanın Yeni Alanları, İstanbul, Devin Yayıncılık, 2005.

—. Resource Wars: The New Landscape of Global Conflict, New York, Metropolitan Books Henry Holt, 2002.

—. Rising Powers Shrinking Planet-The New Geopolitics of Energy, New York, Metropolitan Books. Henry Holt and Company, 2008.

Klare, Michael, Blood and Oil-The Dangers and Consequences of America's Growing Petroleum Dependency, London, Penguin Books Ltd, 2004.

Lovins, Amory B., Energy Security Facts, USA, Rocky Mountain Institute, 2003.

Pamir Necdet, Oğuz Türkyılmaz, "Türkiye`nin Doğalgaz Temin ve İhrac Seçenekleri", Uluslararası Doğalgaz Kongresi ve Sergisi Konferans Tutanakları, İstanbul, 2007.

Pamir, Necdet, "Enerji Arz Güvenliği ve Türkiye", Stratejik Analiz, Mart 2007, pp. 14-24.

Peters, Susanne, "Coercive Western Energy Security Strategies: 'Resource Wars' as a New Threat to Global Security", Geopolitics, Vol. IX, No: 1, Winter 2004, pp. 187-212.

Rees, Judith, Natural Resources, UK, Methwen&Co. Ltd, 1985.
Reuveny, Rafael, "Economic Growth, Environmental Scarcity, and Conflict", Global Environmental Politics, Vol. II, No: 1, 2002, pp. 83-110.
Roberts, John M., "The Black Sea and European Energy Security", Southeast European and Black Sea Studies, Vol. VI, No: 2, 2006, pp. 207-223.
Roberts, John, Pipeline Politics: The Caspian and Global Energy Security, London, Chatham House Publishers, 2007.
Sönmezoğlu, Faruk, Uluslararası Politika ve Dış Politika Analizi, İstanbul, Der Yayınları, 2005.
Sönmezoğlu, Faruk, Uluslararası Politikada Yeni Alanlar Yeni Bakışlar, İstanbul, Der Yayınları, 1998.
Tabb, William K., "Resource Wars", Monthly Review, Vol. LVIII, No: 8, 2007, pp. 32-42.
The World Bank Group, Energy Security Issues, World Bank Working Paper Series, Washington D.C., December 2005, Report No. 36110.
The World Commission on Environment and Development, Our Common Future, London, New York, Oxford University Press, 1987.
Uğurlu, Örgen, "Türkiye'nin Enerji Güvenliğini Yeniden Tanımlamak", 7. Ulusal Çevre Mühendisliği Kongresi Bildirileri, İzmir, Ekim 2007.
Uğurlu, Örgen, Çevresel Güvenlik ve Türkiye'de Enerji Politikaları, İstanbul, Örgün Yayınevi, 2009.
Valdes, J.A. Tapia, "A Typology of National Security Policies", The Yale Journal of World Public Order, Vol. IX, No: 10, 1982, pp. 10-39.
Verrastro, Frank, Sarah Ladislaw, "Providing Energy Security in an Interdependent World", The Washington Quarterly, Vol. XXX, No: 4, 1997, pp. 95-104.
Westing, Arthur H., Global Resources and International Conflicts: Environmental Factors in Strategic Policy and Action, London, Oxford Presss, 1986.
Winrow, Gareth M., "Energy Security in the Black Sea Region: Economic Interdependence or Commercial and Political Rivalry?", Southeast European and Black Sea Studies, Vol. II, No: 2, May 2002, pp. 129-152.
—. "Turkey and the East-West Gas Transportation Corridor", Turkish Studies, Vol. V, No: 2, 2004, pp. 23-42.
Winrow, Gareth, "Geopolitics and Energy Security in the Wider Black Sea Region", Southeast European and Black Sea Studies, Vol. VII, No: 2, 2007, pp. 217-235.

Yergin, Daniel, "Ensuring Energy Security", Foreign Affairs, Vol. 85, Issue: 2, March-April 2006, pp. 69-82.

# CONCLUSION

# FILIZ KATMAN[1]

"Climate Summit 2014-Catalyzing Action" held by UN on 29 September 2014 reiterated once more that climate change in not "a far-off problem". On 12 November 2014, two giant energy emmitters, namely the USA and China agreed to cooperate on curbing greenhouse gas emmissions. Agreement between USA and China on 12 November 2014 is a new hope towards 2015 Paris climate negotiations. USA aims to achieve a target of reducing its emissions by 26 percent to 28 percent in 2025, as compared to 2005 while China aims to peak carbon emissions by 2030 and to have non-fossil fuels by 20 percent in primary energy consumption by 2030. Despite initially ambitious initiatives, the issue of climate change has been pushed down the agenda of world politics. There are seveal reasons for this. According to a special report published in June 2013 by the International Energy Agency, and entitled 'Redrawing the Energy-Climate Map',[2] the target of limiting the long-term rise in the average global temperature by 2°C should be raised to between 3.6°C and 5.3°C, with most of the increase occuring this century. According to the report, carbon-dioxide levels in the atmosphere are exceeding 400 parts per million for the first time in several hundred millennia; also, global energy-related emissions increased by 1.4% to reach 31.6 gigatones (Gt) in 2012, a historic high. This means that alarm bells are ringing. 60% of emissions are now produced by non-OECD countries, a major increase from 45% in 2000. On the other hand, according to the report, due to the deployment of renewables and a significant improvement in the energy intensity of its economy, China's contribution to global emissions is the largest, with the

---

[1] Assistant Professor, Department of Political Science and International Relations, Faculty of Economics and Administrative Sciences, Executive Director, Energy Politics and Markets Research Center, Istanbul Aydin University, TURKEY, filizkatman@aydin.edu.tr

[2] For the report International Energy Agency, " Redrawing The Energy-Climate Map", World Energy Outlook Special Report, 10 June 2013 http://www.worldenergyoutlook.org/media/weowebsite/2013/energyclimatemap/RedrawingEnergyClimateMap.pdf

lowest growth. A similar picture is observed in the US, with a reduction of 200 million tonnes (Mt) in emissions as a result of the switch from coal to gas power generation in the mid-1990s. In Europe, emissions have declined by 50 Mt, due to the economic crisis, growth in renewables and a cap on emissions from the industry and power sectors. In another developed country, Japan, the increase in emissions amounts to 70 Mt. This is due to efforts at improving energy efficiency not fully offsetting fossil fuel use, in order to compensate for the reduction in nuclear power in the wake of Fukushima.

As the figures in several countries show, the realization of the target in limiting the long-term rise in the average global temperature by 2°C is heavily dependent on energy, since two-thirds of emissions occur in the energy sector, and more than 80% of global energy consumption is based on fossil fuels. A realistic picture reveals the fact that global energy-related greenhouse gas emissions in 2020 are projected to be nearly 4 Gt $CO_2$-equivalent ($CO_2$-eq) higher than the level that would be consistent with attaining the 2°C target. This is so even after allowing for the policies now being pursued. This highlights the scale of the challenge still to be tackled in this decade alone.

The report also recommends policies needed to keep the door open to the 2°C target through to 2020, at no net economic cost. These recommended policies aim to reduce greenhouse-gas emissions by 3.1 Gt $CO_2$-eq in 2020 (80% of the emissions reduction required under a 2°C trajectory), and to provide time while international climate negotiations continue in the lead up to the important Conference of the Parties meeting in Paris in 2015, and national policies necessary for implementing the expected international agreement are put in place. It also mentions that the policies should meet key criteria, such as delivering significant reductions in energy-sector emissions by 2020 (as a bridge to further action), relying only on existing technologies, being already adopted and proven in several countries and not harming economic growth in any country or region as a result of their widespread and collective adoption. The four policies are:

- Adopting specific energy efficiency measures (49% of emissions savings).
- Limiting the construction and use of the least-efficient coal-fired power plants (21%).
- Minimising methane ($CH_4$) emissions from upstream oil and gas production (18%).
- Accelerating the (partial) phase-out of subsidies to fossil-fuel consumption (12%).

In the report, the interdependence between global climate and the energy sector is identified in the sudden and destructive effects (caused by extreme weather events) which pose risks to power plants and grids, oil and gas installations, wind farms and other infrastructures, and also more gradual impacts, such as changes to heating and cooling demand, sea level rise on coastal infrastructure, shifting weather patterns on hydropower and water scarcity on power plants, and knock-on effects on other critical services due to disruptions to the energy system. Responsibility in managing the process is given to the government, which should design and implement frameworks to encourage adaptation, and to the private sector, which should assess risks and impacts as part of its investment decisions.

While such policy recommendations are proposed, there remains a significant obstacle to effectively tackling climate change, namely the issue of international climate negotiations. The key document, the Kyoto Protocol, is far from reliable, since it is off the primary track and has even became meaningless. Today, with rapidly worsening figures relating to global climate change, the self-maximisation attempts of each country (specifically countries such as USA, China, India, Russia, Canada, Australia, and New Zealand) will cause the collapse of the world's eco system not in the long or medium term, but in the short run; this is a tragic game, in which everyone loses.

USA, China, India, Russia, Canada, Australia and New Zealand are the countries responsible for roughly 80% of greenhouse gas emissions. China and India are not present in the Kyoto Protocol in any meaningful way; along with the USA, which did not participate in the first obligation period, Russia, Canada and Japan announced their withdrawal from the second obligation period. It would be over-optimistic to expect a contribution to a solution from the participation in the second obligation period of only the EU and Norway, which control only about 20% of global greenhouse gas emissions. Here, specifically, the attitude of the USA, China, India, Russia and Canada is crucial. The protocol is subject to the participation of these countries, primarily the USA, as the world's biggest emitter of $CO_2$. The collective struggle against climate change requires that the primary actors in greenhouse gas emissions caused by human activities take responsibility. Despite the fact that these countries benefit from the earth's resources, the wrong decision now will lead to the absolute failure of such policies and future generations will be harmed if these countries fail to acknowledge the economic, social, environmental costs of the struggle against global climate change. World leaders invited by UN General Secretary Ban Ki-Moon to Climate Change 2014 Summit are asked "to make bold announcements and actions that will reduce emissions, strengthen climate resilience and mobilize political will for a legal agreement in 2015". It is stated that "Climate Summit 2014 provides a unique opportunity for leaders to champion an ambitious vision, anchored in action that will enable a meaningful global agreement in 2015".

In this light, in the first chapter of the book, issues of global climate, environment and energy are discussed in three articles focusing on global challenges and opportunities to global stability, with a primary focus on the security aspects of global climate change. In the article entitled "Analysis of Global Climate Change Impacts on Security" by Filiz Katman, it is argued that global climate change will worsen conditions on earth in terms of air quality, plant quality, food quality, water quality, natural disasters and energy resources, which all serve for the survival of human beings, the main element of security. So, in relation to global climate change and security, the primary concern will be survival. Global climate change poses a serious threat to survival, the main goal of human beings. Due to the direct relationship between threat and security, global climate change has become one of the most serious threats to security, and requires the implementation of necessary precautions in order to protect from the serious impacts of global climate change. Not only the environment but also the economy will be damaged by climate change. Deteriorations in living conditions on earth will threaten the possibility of sustainable economic growth in every country. Raising awareness, creating public pressure on governments, designing targets and pricing carbon emissions are the steps which are lacking in managing the impacts of climate change.

The article entitled "Paradigm Shift in Millennium Development Goals Management through Global Climate" by Aylin Met and Taner Altunok argues that analyses performed in the year 2010, 5 years before the 2015 deadline for reaching the Millennium Development Goals (MDGs), have shown that it became much harder to achieve MDGs compared to when they were first accepted, and that this is due to the financial, food and climate change crises world has faced. It is foreseen that climate change could affect MDGs in a couple of ways. The OECD's 2010 report indicates the first steps in a paradigm shift in MDG management through climate change, with the phrase "development as usual is no longer an option." It is obvious that responsibilities for the achievement of MDGs are not possible through a circumspect development agenda.

The article entitled "Seeking Security in Changing Climate" by Argiro Vasilaki argues that on 11 February 2011, the 6479th meeting of the United Nations Security Council stressed the need to consider economic and social—as well as political—factors in maintaining international peace and security. The effects of climate change were addressed as factors that are having an increasingly serious impact on development and security. Climate change is viewed as a "threat multiplier" by the United Nations and the European Union. Food insecurity and water shortages are challenging the basis of human life. Desertification, the loss of coastal zones and natural disasters, combined with ineffective forms of adaptation and mitigation, could lead to increasing numbers of environmentally displaced people.

Chapter Two is dedicated to a discussion of the impacts of Global Climate Change on Environment, and includes six articles by scholars adopting a multidisciplinary approach. The article entitled "A New Ethical Approach to Environmental Protection: Ecocentrism", by Kıvılcım Akkoyunlu Ertan and Birol Ertan, argues that Turkey, which issued trade freedom in 1980 and fiscal freedom in 1989, will not escape the present crises, despite the economic policy changes it has implemented in order to prevent them. As is the case all around the world, these crises were experienced in Turkey after financial liberalisation.

The article entitled "Voluntary Carbon Trade: A New Livelihood Model for Low Income Families", by İlker Met and Taner Altunok, argues that the "emission trade" or "carbon trade" used to control greenhouse gases has developed and become an important part of the agenda since the effects of global climate change have become more obvious. As a result, the size of the carbon market, which was 20 billion dollars in the year 2006, reached 143.7 billion dollars by 2009, in spite of the 2008 global financial crisis. This volume is an indicator of the importance of the subject. Although until now, decreasing carbon levels seems to be achievable only by means of high budget projects, many projects which were just a dream a decade ago have made it possible to consider the carbon trade as an opportunity for low income families. In fact, several carbon projects providing employment opportunities for poor households and thus for national development have been developed globally in the last two years. But the necessary standards and legal frameworks have not yet been completed. When one considers that many low income families do not own homes or land, it is clear that the question of how poor households will take advantage of the carbon trade will become an important topic in the near future. The chapter presents practices of the carbon market appropriate for low income families (and presents a resulting model for Turkey). It is argued that that this represents not only a way of saving low income families, who are the most vulnerable to the effects of climate change, but also an important step towards mitigating the negative effects of climate change, by liberating the carbon market from the monopoly of large projects and making it available to the community.

The article entitled "Analysis of Turkish Environmental Energy Policies in a Regional Framework" by Çiğdem Üstün argues that climate change, a decrease in biological diversity, concerns over water resources, and increasing individual and industrial energy needs deeply affect international and national energy and environmental policies. Today, states not only find it necessary to secure energy supplies but also to address these environmental matters. Turkey also feels these necessities and requirements, and therefore its evolving relations with neighbours focus on environment and energy policies. Issues such as droughts and other decreases in water resources and the transit passage of oil tankers through the Turkish Straits force Turkey to emphasise the production and use of environmentally friendly energy. In particular, oil tanker traffic in the Black Sea has a negative impact on biological diversity. However, as Turkey makes agreements with the energy producing countries for the construction of various pipelines, Turkey's concern has not only been the potentially hazardous environmental situations, but also the economic and geopolitical situation of the country. Unfortunately, this understanding decreases the usage of renewable energy resources in Turkey.

In the article entitled "European Union Environmental Directives and a Practice on Business," Meral Erol Fidan argues that environment is one of the most problematic issues for Turkey during its process of adaptation to the European Union. About 300 directives of the *acquis communitaire* are directly related with various environmental issues. Being a candidate for full membership, Turkey's harmonisation efforts with EU directives are still continuing. However, it is estimated that businesses in Turkey do not have enough knowledge about environment and waste management and do not put sufficient effort into this field. The existing situation in Turkey and efforts to measure the level of the businesses' knowledge about environment and waste management as part of the process of adapting to the EU are analysed in small and medium sized businesses, both inside and outside the IZs in Sakarya, one of the Turkey's most important industrial cities. The data from 112 enterprises are analysed using SPSS 16.0 for Windows, and the current situation and the differences between enterprises of various kinds is determined.

The article entitled "Wind Speed and Energy Characteristics based on Statistical Analyses" by Zafer Aslan and Z. Nevin Çağlar is related to national energy plans on wind energy converting systems (WECSs). An expected energy output for different sizes is calculated. It is recommended that one large size wind turbine would facilitate the construction of a water pumping system in the vicinity of Şile. These results are similar to those obtained via energy potential analyses in some regions of the Aegean Sea, and will contribute to decision-making and planning national renewable energy processes.

Chapter Three is based on a discussion concerning the impacts of global climate change on energy. The article entitled "Sustainable Development and Turkey's Biomass Energy Potential" by Coşkun Karaca and M. Mustafa Erdoğdu argues that achieving sustainable development depends on good governance practices; in particular, on the effective implementation of national sustainable development strategies which integrate government decision-making within the economic, environmental and social spheres. To achieve a rapid increase in biofuels production that can be sustained over the long term, policies that are consistent, long term, and supported by broad stakeholder participation are needed.

The article entitled "Energy and Climate Change Policies of Russia and China" by Çağla Gül Yesevi and Ceren Uysal Oğuz argues that the Russian Federation and China are both key actors in terms of global energy supply and GHG emissions. For that reason, their active participation is indispensable to an effective climate change mitigation and adaptation structure, especially in the post-Kyoto period. Therefore, it is necessary to examine the energy and climate change policies of these countries, in order to evaluate their contribution to the international climate regime and the importance of their energy strategies in international relations.

The article entitled "Energy Security Concept in Global Climate Change Process and Defining the Concept with Economic, Political and Environmental Aspects" by Hakan Korhan argues that the development of the energy security concept, especially in the 2000s, gained a place beyond natural resources security. The existence of different dimensions of the concept of energy security renders the definition of the concept difficult, and at the same time this situation shows the theoretical limitations of economics, political science and theories of international relations in analysing the concept of energy security. This, in turn, constrains the establishment of a theoretical framework and gives rise to the obligation to define the concept within a multi-disciplinary approach. The energy security policies of countries may vary depending on the emphasis put on different dimensions of the concept, and this causes the meaning given to energy security by each country to be different. It is possible to analyse energy security policies in terms of different purpose categories in foreign policy. In accordance with the increasing importance of the matter, the place of energy security has been rising in the rankings made by countries relating to their foreign policy purposes.

# CONTRIBUTORS

**Taner Altunok,** Professor of Mechanical Engineering, Director, Graduate School of Natural and Applied Sciences, Çankaya University

**Zafer Aslan,** Professor of Computer Engineering, Faculty of Engineering and Architecture, Istanbul Aydin University

**Z. Nevin Çağlar,** Engineer, Meteorology Laboratory, Kandilli Observatory and Earthquake Research Institute, Bosphorus University

**M. Mustafa Erdoğdu,** Professor, Department of Public Finance, Faculty of Economics, Marmara University

**Birol Ertan,** Associate Professor, Lecturer, TODAİE

**Kıvılcım Akkyounlu Ertan,** Associate Professor, Lecturer, TODAİE

**Merol Erol Fidan,** Assistant Professor, Bilecik Şeyh Edebali University

**Coşkun Karaca,** Assistant Professor, Department of Public Finance, Faculty of Economics and Administrative Sciences, Cumhuriyet University

**Filiz Katman,** Assistant Professor, Department of Political Science and International Relations, Faculty of Economics and Administrative Sciences; Executive Board President, Energy Politics and Markets Research Center, Istanbul Aydin University

**Hakan Korhan,** PhD, Turkey Representative, Credit Europe Bank NV

**Aylin Met,** PhD, Scientific Programs Specialist, TÜBİTAK

**İlker Met,** PhD, Project Manager, Corporate Management Directorate, TR Ziraat Bank

**Ceren Uysal Oğuz,** Assistant Professor, Department of International Relations, Faculty of Economics and Administrative Sciences, Istanbul Kültür University

**Çiğdem Üstün,** Associate Professor, Chair, Political Science and Public Administration Department, Vice Dean, Faculty of Economics and Administrative Sciences, Gediz University

**Argiro Vasilaki,** Economist, PhD Student, University of the Aegean

**Çağla Gül Yesevi,** Assistant Professor, Department of International Relations, Faculty of Economics and Administrative Sciences, Istanbul Kültür University

# INDEX